Induced current density in human body models caused by inhomogeneous magnetic fields of electrical appliances

Von der Fakultät Elektrotechnik und Informationstechnik an der
Universität Stuttgart zur Erlangung der Würde eines
Doktors der Ingenieurwissenschaften (Dr.-Ing.) genehmigte Abhandlung

vogelegt von

Shinichiro Nishizawa, Ph.D.

geboren in Tokio / Japan

Hauptberichter: Prof. Dr.-Ing. habil. F.M. Landstorfer

Mitberichter: Prof. Dr. techn. W.M. Rucker

Tag der Einreichung: 4. Mai 2005

Tag der mündlichen Prüfung: 13. Juli 2005

Institut für Hochfrequenztechnik der Universität Stuttgart

2005

Bibliografische Information Der Deutschen Bibliothek

Die Deutsche Bibliothek verzeichnet diese Publikation in der Deutschen
Nationalbibliografie; detaillierte bibliografische Daten sind im Internet über
http://dnb.ddb.de abrufbar.

ISBN 3-8325-0997-6

Logos Verlag Berlin
Comeniushof, Gubener Str. 47,
10243 Berlin
Tel.: +49 030 42 85 10 90
Fax: +49 030 42 85 10 92
INTERNET: http://www.logos-verlag.de

Acknowledgements

The present study is pursued at the Institute of Radio Frequency Technology (Institut für Hochfrequenztechnik (IHF)), Department of Computer Science, Electrical Engineering and Information Technology, University of Stuttgart, Stuttgart, Germany.

First, I would like to thank my supervisor,

Prof. Dr.-Ing. habil. Friedrich M. Landstorfer

the head of the institute, for accepting me and guiding the research work that led to this thesis. I would like to express my gratitude for all his contributions. Furthermore, I would like to thank my co-supervisor,

Prof. Dr. techn. Wolfgang M. Rucker

the second joint referee, for his unstinting support of this study. I would also like to thank Dr.-Ing. Hans -Oliver Ruoß, who is a representative researcher for the EMVU group members (IHF & Robert Bosch GmbH), for introducing me to Germany and making my stay to a successful one.

Furthermore, I would like to express my gratitude to:

Dr.-Ing. Wolfgang Spreizer for his cooperation and assistance at the IHF and Bosch, and for guiding me in the field of low frequencies.

PD Dr.-Ing. Ningyan Zhu with whom I had numerous discussions on theoretical problems concerning this study.

Prof. Dr. Osamu Hashimoto (Aoyama Gakuin University, Tokio Japan), who was my supervisor when I was a Ph.D. student in Japan for guiding in the achievement of my first doctoral degree and for his encouragement and support during my stay in Germany.

Prof. Dr. Yoshitsugu Kamimura (Utsunomiya University, Utsunomiya-shi Japan) for prolifically international cooperation work (Joint research between DFG[1] and JSPS[2]), and also performing the dosimetric calculation in this study.

---

[1] DFG: Deutsche Forschungsgemeinschaft
[2] JSPS: Japan Society of the Promotion of Science

IV

Dipl.-Ing. Marcus Maier and Dipl.-Ing. Dirk Zimmermann, who frequently helped me to improve the language (English and German) of the papers.

My ex-[3] and present[4] colleagues for their friendship and kindness during my daily life at the institute. Furthermore, the excursions to south Tirol by the ex- and present members of IHF, which takes place every year, were extremely memorable; the mountain climbing was especially refreshing.

Dipl.-Ing. S. Meßy, N. Angwafo, and L. Paltre, who closely worked with me as diploma students and are presently pursuing their doctoral degree.

For financial support, I am grateful to the Alexander von Humboldt Foundation, Bonn, Germany, and also partly to the JSPS (Postdoctoral Fellowship), Tokio, Japan. Furthermore, I would like to thank the Robert Bosch GmbH (Dr.-Ing. W. Hiller, Dr.-Ing. M. Klauda, Dr.-Ing. T. Ruf, Dipl.-Ing.(FH) E. Fauser, and Dipl.-Ing. A. Quett), Stuttgart, Germany, and Bosch Siemens Haushaltsgerät GmbH (Dipl.-Ing. U. Kampet), Berlin, Germany, for the liberality in granting permission to use the measurement facility.

Finally, I wish to thank the following people: my parents, brother, uncle, aunt, and cousin for their encouragement and support over the years, which allowed me to fully concentrate on my research and thus significantly contribute to the successful completion of this thesis.

---

[3]PD Dr.-Ing. Ulrich Jakobus, Dr.-Ing. J. Waldmann, M. Klar, R. Hoppe, T. Binzer, N. Berger, M. Schick, J.Kantz, L. Geisbusch, W. Spreitzer, H.-O. Ruoß

[4]PD Dr.-Ing. Ningyan Zhu, Dr.-Ing. W. Mahler, Dipl.-Ing. J. Baumann, M. Maier, M. Layh, T. Hiegler, A. Friederich, P. Wertz, M. Leibfritz, D. Zimmermann

Der Intellekt
hat ein scharfes Auge
für Methoden und Werkzeuge,
aber er ist blind
gegen Ziele und Werte.

Intellect has a keen eye
for method and technique
but is blind
to aim and value.

知性は手段および技術に
鋭敏な目を有するが、
目標および目的には盲目だ。

Albert Einstein

VI

Contents

Notations

Abbreviations

EM	*Electromagnetic*
ELF	*Extremely Low Frequency*
ICNIRP	*International Commission on Non-Ionizing Radiation Protection*
CENELEC	*European Committee for Electrotechnical Standardization*
EN	*European Norm*
CEN	*European Committee for Standardization*
SPFD	*Scalar Potential Finite Difference*
SOR	*Successive Over-Relaxation*
SAR	*Specific Absorption Rate*
IH	*Induction Heater*
DNA	*Deoxyribonucleic Acid*

Symbols

div		Divergence operator
grad		Gradient operator
rot		Rotation operator (curl)
$\Diamond$		Vector laplace operator
ρ	As/m^3(C/m^3)	Electric charge density
ϵ_0	As/Vm	Permittivity in free space
ϵ_r		Relative permittivity
μ_0	Vs/Am	Permeability in free space
μ_r		Relative permeability
σ	S/m	Conductivity
λ_0	m	Wavelength in free space
f	Hz	Frequency
ω	1/s	Angular frequency
$\vec{A}$	Vs/m(Tm)	Vector potential
ψ		Scalar potential
$\vec{J}_*$	A/m	Surface electric current
$\vec{M}_*$	V/m	Surface magnetic current
$\hat{n}$		Unit vector
$\vec{B}$	Vs/m^2(T)	Magnetic flux density
$\vec{H}$	A/m	Magnetic field
$\vec{D}$	As/m^2(C/m^2)	Electric flux density
$\vec{E}$	V/m	Electric field
$\vec{J}$	A	Current density induced in the biological tissues
r	m	Distance from the hot spot (EN50366 [5])
r_{coil}	m	Radius of the coil model (EN50366 [5])
d_{coil}	m	Distance of the coil model (EN50366 [5])
d_{cyl}	m	Distance between the equivalent source model and the appliance surface
G	m	Integrated value of measured magnetic flux density (EN50366 [5])
$\vec{m}$	Vs m	Magnetic dipole moment
δ	m	Skin depth
d	m	Distance between the source and the body model

Zusammenfassung

Mit der weltweiten Verbreitung von Mobilfunktelefonen steigt in der Öffentlichkeit die Besorgnis über die gesundheitlichen Auswirkungen elektromagnetischer Felder, einschließlich der Störfelder, die von elektrischen Haushaltsgeräten und anderen Elektrogeräten ausgehen. In internationalen Gremien wurden zum Schutz der Benutzer vor diesen möglicherweise hohen elektromagnetischen Feldern Grenzwerte festgelegt. Um eine mögliche Gefährdung von Personen mit Hilfe numerischer Methoden entsprechend bewerten zu können, müssen die magnetischen Störfelder von Geräten numerisch nachgebildet werden. Die vorliegende Arbeit befasst sich mit einem neuen numerischen Modell, dem Äquivalenten Quellenmodell. Dieses Äquivalente Quellenmodell ermöglicht eine genaue Beschreibung der Exposition des Benutzers im vom Gerät ausgehenden komplexen magnetischen Störfeld (dreidimensionale Darstellung des magnetischen Vektorfeldes), was auch eine genaue numerische Dosimetrie-Berechnung für niedere und mittlere Frequenzbereiche ermöglicht. Im Herbst 2004 wurde das Äquivalente Quellenmodell als standardisiertes Bewertungsverfahren in die europäische Norm EN50366 (CENELEC) eingebracht.

Das Bild auf der Seite XII zeigt, wie die Kapitel in der vorliegenden Arbeit aufeinander aufbauen. Nach einigen einleitenden Bemerkungen in **Kapitel 1** werden in **Kapitel 2** Hintergrundinformationen zu den biologischen Auswirkungen von elektromagnetischen Feldern zusammengestellt. Weiterhin werden einige Theoreme und Grundlagen aus dem Bereich der Feldtheorie vorgestellt, die für diese Arbeit erforderlich sind. Es findet sich eine Einführung in die Maxwell'schen Gleichungen und die Potenziale, die bei den grundlegenden Gleichungen für die SPFD-Methode (Scalar Potential Finite Difference Method) eingesetzt werden. Außerdem wird das Äquivalente Quellenmodell vorgestellt und die Einzelheiten dieses Modells erklärt. Darüber hinaus enthält dieses Kapitel eine Erörterung des grundlegenden Äquivalenzprinzips, das in dem neuen Quellenmodell Anwendung findet. Im letzten Teil dieses Kapitels werden die Grundprinzipien der numerischen Methode (SPFD-Methode) erläutert, die bei der dosimetrischen Berechnung verwendet wird.

Kapitel 3 enthält eine Beschreibung des Äquivalenten Quellenmodells und der inhomogenen Magnetfelder. Im ersten Teil des Kapitels wird die Plausibilität des vorgeschlagenen neuen Quellenmodells numerisch geprüft, indem die Genauigkeit der Magnetfeldverteilung für das numerische Referenzmodell einer Stromschleife untersucht wird. Die ermittelten Werte für magnetische Feldstärke, Feldvektoren (Polarisation) und die Richtung der Feldintensität (Strahlrichtung) waren bei beiden Modellen annähernd deckungsgleich. Diese hohe Übereinstimmung stellt eine Bestätigung der numerischen Plausibilität des Äquivalenten Quellenmodells dar. Außerdem wurde die Plausibilität auch für handelsübliche Geräte ermittelt, indem die Ergebnisse der gemessenen und der berechneten magnetischen Flussdichte für die Geräte mit Hilfe eines 3D-Scan-Messgeräts verglichen wurden. Die gemessenen und die berechneten Ergebnisse zeigen eine hohe Übereinstimmung, und somit ist die Plausibilität und die praktische Anwendbarkeit des Äquivalenten Quellenmodells auch für reale Geräte bestätigt.

Im zweiten Teil von Kapitel 3 wird das validierte Äquivalente Quellenmodell benutzt, um die Eigenschaften des magnetischen Störfeldes in der Umgebung von Haushaltsgeräten zu untersuchen. Bei diesen Untersuchungen wurden die unterschiedlichen Feldeigenschaften geprüft, die sich aus dem verschiedenen internen Aufbau der Geräte ergeben. Die Genauigkeit der magnetischen Feldeigenschaften wird anhand des Spulenmodells, das gemäß der europäischen Norm EN50366 vorgeschrieben ist, untersucht. Eine hohe Übereinstimmung zwischen dem Spulenmodell und dem Testgerät wird für die magnetische Feldstärke auf der Achse bestätigt, die die gleiche Richtung wie die Feldintensität (Strahlrichtung) des Spulenmodells hat. Allerdings ist beim dominanten Feldvektor eine erhebliche Differenz zwischen dem Spulenmodell und dem Induktionskochherd (einem der Testgeräte) zu beobachten. Der Grund dafür ist der Unterschied zwischen der Anordnung der Heizspirale, die sich unter dem Topf im Gehäuse des Induktionskochherds befindet und dem Spulenmodell.

Die dosimetrischen Untersuchungen werden in **Kapitel 4** beschrieben. Die induzierte Stromdichte in den numerischen Körpermodellen wird mit Hilfe der SPFD-Methode bei niedrigen und mittleren Frequenzen (50 Hz bzw. 21 kHz) numerisch berechnet. Im ersten Teil dieses Kapitels werden die dosimetrischen Unterschiede zwischen den numerischen Körpermodellen anhand von handelsüblichen Haushaltsgeräten untersucht. Bei den numerischen Modellen werden für diese Untersuchung das anatomisch realistische Körpermodell und die vereinfachten Körpermodelle mit homogener Leitfähigkeit verwendet. Die Stromdichten, die sich daraus ergeben, werden mit den internationalen Grenzwerten (ICNIRP Guideline) verglichen und anhand des anatomischen Körpermodells bestätigt, so dass die maximalen Stromdichten innerhalb dieser Grenzwerte für eine typische Expositionsanordnung liegen, auch wenn der Wert der magnetischen Flussdichte die Referenzwerte überschreitet. Außerdem haben die Ergebnisse gezeigt, dass die maximalen Stromdichten bei den homogenen Körpermodellen an der Oberfläche auftreten, beim anatomischen Körpermodell dagegen im Körperinneren.

Ähnlich wie im zweiten Teil von Kapitel 3 wird die Genauigkeit der induzierten Stromdichte für das Spulenmodell untersucht, das in der EN50366 vorgeschrieben ist. Bei dieser Untersuchung wird die maximale Stromdichte für das Spulenmodell und die Testgeräte geprüft. Diese Ergebnisse haben gezeigt, dass die Stromdichte mit dem Spulenmodell dreimal so hoch sein kann wie mit dem tatsächlichen Gerät. Dies ist auf Unterschiede beim dominanten Feldvektor zurückzuführen. Schließlich werden in **Kapitel 5** die in dieser Arbeit vorgestellten Ergebnisse zusammengefasst.

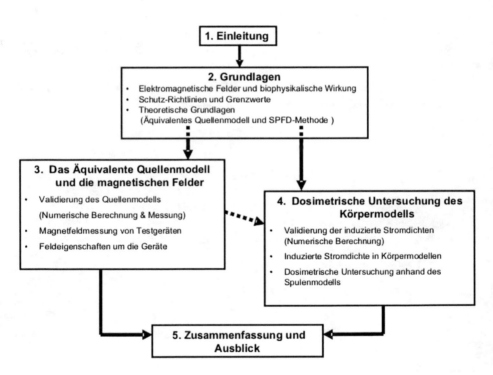

Abbildung: Kapitelablauf über diese Dissertation.

Summary

With the global proliferation of mobile phones, the public has become increasingly concerned about the effects of electromagnetic fields on health, including the fields emitted by electrical household appliances and other electronic devices. International bodies have standardized the exposure limits to safeguard users from these possibly high electromagnetic fields. In order to appropriately evaluate possible exposure hazards by numerical methods, it is necessary to accurately describe the magnetic field associated with the electrical appliances. This thesis deals with a new numerical model (equivalent source model). This model enables the reproduction of complicated inhomogeneous magnetic field characteristics associated with practical electrical appliances with a complete three-dimensional representation of the magnetic field vectors (field components and directions) for the purpose of validating the inhomogeneous magnetic field dosimetry at low and intermediate frequency ranges. In the autumn of 2004, this proposed new source model has been adopted as one of the standard evaluation methods in the EN50366 (CENELEC).

The figure on page XV shows the development flow of the chapters in this thesis and begins with some introductory remarks in **Chapter 1**. The following **Chapter 2** summarizes the background information of the biological effects from electromagnetic field radiation and the basics of the field theory, which is limited to the discussions necessary for this work. An introduction to Maxwell's equations and the potentials that are applied in the basic equations of the scalar potential finite difference (SPFD) method are described in this chapter. Furthermore, the equivalent source model is proposed and the details of this source model are described. Fundamental theory such as the equivalence principle, which is applied in the proposed new source model, is also discussed. The basic principles of the numerical method (SPFD method), which is used in the dosimetric calculation, are explained in the final part of this chapter.

Chapter 3 describes the equivalent source model and the inhomogeneous magnetic fields. In the first part of this chapter, the validity of the proposed new source model is numerically investigated by evaluating the accuracy of the magnetic field distribution around the geometry-based numerical reference model of a current loop. Approximately identical magnetic field strengths, directions of the field strength vectors (polarization), and directions of the field intensity (main beam) are obtained for both models. Due to this good agreement, the validity of the equivalent source model is numerically confirmed. Furthermore, its validity is also confirmed for commercial appliances by comparing the results of the measured and calculated magnetic flux density associated with the appliances using the measurement equipment "3D-scan". The measured and calculated results are in good agreement, and the validity of the equivalent source model is confirmed for real household appliances (i.e. practical application). In the second part, the validated new source model (i.e. equivalent source model) as described above is applied to investigate the magnetic field properties associated with various types of household appliances; this investigation considers the difference in the nature of the various magnetic fields resulting from the different inner construction. The accuracy of the magnetic field properties is investigated using the coil model, which is prescribed in the European norm EN50366. A good agreement between the coil model and the test appliance is confirmed for the magnetic field strength on the axis, which has the same direction as the field intensity calculated with the coil model. However, a significant difference is observed in the dominant field strength vector between the coil model and the induction heater, which is used as a test appliance. This is caused by the difference in the arrangement of the eddy heating coil (located under the vessel in the induction heater body) and the coil model.

The dosimetric studies are described in **Chapter 4**. The induced current density in the numerical body models are calculated using the SPFD method at low (50 Hz) and intermediate (21 kHz) frequencies. In the first part of this chapter, the dosimetric differences between the different numerical human body models are investigated using commercial household appliances. In the case of the numerical human body models, the anatomically realistic one and the simple human body models (with homogeneous conductivity) are used for this investigation. The resulting current densities are compared with the international limits (basic restriction), and are confirmed by applying the anatomical body model that the maximum values of the induced current density satisfy these limits for a typical arrangement position of the appliance, even if the magnetic flux density value exceeds its limits (reference level). Furthermore, it is observed that the maximum induced current density appears on the surface with the homogeneous body models, while it appears inside the body model for the anatomically realistic case. Similar to the second part of Chapter 3, the accuracy of the induced current density is investigated for the coil model, prescribed in EN50366. In this study, the maximum current density induced in the body models is investigated for the coil model and the test appliances. These results reveal that the value observed for the coil model could be three times that of the real appliance; this is caused by differences in the dominant field strength vector. Finally, **Chapter 5** summarizes the results presented in this work.

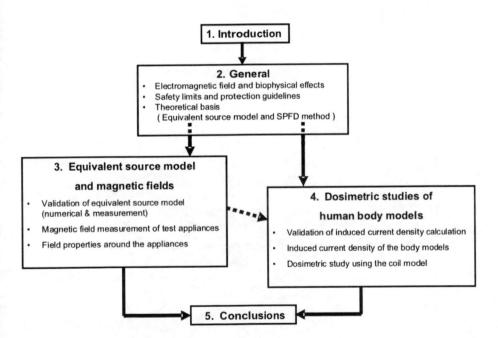

Figure: Chapter flow of this thesis.

Chapter 1

Introduction

International bodies have set up guidelines and standards for exposure limits [1] for protecting users of electric and electronic devices (e.g. household appliances, electric shavers, hair dryers, etc. [2]) against the emission of non-ionizing low-frequency magnetic fields. Furthermore, the European Commission has requested CEN and CENELEC for new European standards, which describe the measurement and calculation methods in order to specify product requirements, limiting the emission of electromagnetic fields from the electric appliances ([3],[4]). Currently, a new standard measurement and calculation method (EN50366 [5]) for household appliances (up to 400 kHz) has been enforced by the European committee for electrotechnical standardization (CENELEC). In order to appropriately evaluate the possible exposure hazards by numerical methods, it is necessary to describe the magnetic field around the electrical appliances accurately. However, a complete geometry-based numerical modeling of a real magnetic source (e.g. the motor of the electrical appliance) is enormously complex and, therefore, not applicable to the intended dosimetric investigation. Consequently, a current loop [6], a multiturn current solenoid [7] and a magnetic dipole moment ([8],[9],[10]) have been commonly used as simple numerical source models, which locally generate a magnetic field strength equivalent to the values measured around the device. With these simple source models, it is possible to postulate the matching of the magnetic field components for one single axis direction, but not for all directions (three-dimensional). In the case of dosimetric investigations, the direction of the magnetic field intensity and the polarization, which varies extremely with the structure and arrangement of the real magnetic sources (e.g. motor) inside an electrical appliance, is of interest. However, the accurate treatment of such magnetic field characteristics is important for an accurate dosimetric evaluation [12].

Hence, a new numerical source model (equivalent source model) is proposed and validated in this thesis. This new source model allows the reproduction of the complicated inhomogeneous magnetic field characteristics associated with real electrical appliances with a complete three-dimensional representation of magnetic field vectors (field components and directions) for the purpose of validating inhomogeneous magnetic field dosimetry[1] at the low and intermediate frequency range. In the autumn of 2004, this proposed source model has been adopted as one of the standard evaluation methods in the European norm EN50366.

By applying this source model, the magnetic field properties and dosimetry are investigated for various types of household appliances; the investigations consider the nature of various magnetic fields caused by different inner constructions. The scalar potential finite difference (SPFD) method is applied for numerical calculations and the dosimetric differences among the numerical human body models are investigated. The anatomically realistic model as well as simple human body models (with homogeneous conductivity), which have already been used in former studies, are investigated. In these studies, the induced current density is also compared with the international standard limits (ICNIRP guideline) and the safety of the test appliances is discussed.

As described above, a new standard evaluation method for the household appliances has been currently enforced by the CENELEC. In the new standard, EN50366 [5], two types of source models are provided, which allow the simulation of the magnetic field distribution around various test appliances and the procedure for its determination is defined in general terms. These two source models are used in different evaluation steps, provided that the magnetic field levels of the household appliances exceed the limit of the reference level given by the ICNIRP guideline [1]. This emphasizes the necessity for assessing compliance with the basic restrictions offered by the ICNIRP guideline. The first evaluation method is a simplified method and enables only an approximated assessment. In this method, a coil model is applied

---

[1]The meaning of dosimetry for electromagnetic field radiation is not similar to that for ionizing radiation. Dosimetry is interpreted as means to determine electromagnetic fields inside the biological body [14].

as the substitute numerical source model instead of the test appliance. The second method is an exact evaluation method one, which is applied when the induced current obtained from the simplified method (coil model) exceeds the basic restrictions. In this method, the equivalent source model described in this thesis is applied. However, the accuracy of the magnetic field and the induced current density, which are simulated with the coil model (defined in EN50366), have not yet been verified. Hence, the accuracy of the magnetic field properties and dosimetry, which are calculated from the coil model, are also investigated.

Chapter 2

General

2.1 Electromagnetic fields and the biophysical effects

2.1.1 Maxwell equations and the scalar-vector potential

All classical electromagnetic phenomena can be summarized by a set of four equations known as the Maxwell equations described below:

$$\text{rot } \vec{E} = -\frac{\partial \vec{B}}{\partial t} \quad \text{(Faradays law of induction)} \tag{2.1}$$

$$\text{rot } \vec{H} = \vec{J} + \frac{\partial \vec{D}}{\partial t} \quad \text{(Ampere − Maxwells law)} \tag{2.2}$$

$$\text{div } \vec{B} = 0 \quad \text{(Absence of free magnetic poles)} \tag{2.3}$$

$$\text{div } \vec{D} = \rho \quad \text{(Gausss law for electric fields)} \tag{2.4}$$

Equation (2.1) presents a differential form of Faradays law of induction. It states that a time-varying magnetic field induces an electric field ($\vec{E}$ field). Equation (2.2) represents a vector form of Ampere-Maxwells law. The second term of this equation is known as the displacement current, an element which was added by Maxwell. The presence of the second term on the right-hand side means that a time dependent $\vec{E}$ field induces a magnetic field $\vec{H}$ in the high frequency range, even without current flow in a conductor. Equation (2.3) indicates that magnetic fields have no point sources where the field lines could begin or end, and means that the magnetic fields are continuous. Equation (2.4) indicates that an $\vec{E}$ field may begin or end on an electric charge density of ρ. It represents Gauss's law for electric fields.

The Maxwell equations consist of a set of coupled first-order partial differential equations regarding the various components of electric and magnetic fields. They can be solved as they stand in straightforward situations. But it is often advisable to introduce potentials, while also satisfying some of the Maxwell equations identically. Furthermore, this potential approach has been applied in the Section 2.2.2, in the basic theory of the numerical means of calculation (SPFD method). Since equation (2.3) still holds, we can define the magnetic flux density $\vec{B}$ in terms of a vector potential $\vec{A}$:

$$\vec{B} = \text{rot } \vec{A}. \tag{2.5}$$

Then the equation (2.1) can be written as:

$$\text{rot}\left(\vec{E} + \frac{\partial \vec{A}}{\partial t}\right) = 0 \tag{2.6}$$

This means that the quantity with vanishing curl in equation (2.6) can be written in terms of the gradient of a scalar function (scalar potential ψ):

$$\left(\vec{E} + \frac{\partial \vec{A}}{\partial t}\right) = -\text{grad } \psi$$

$$\vec{E} = -\text{grad } \psi - \frac{\partial \vec{A}}{\partial t} . \tag{2.7}$$

The definitions of $\vec{B}$ and $\vec{E}$ in terms of potentials $\vec{A}$ and ψ according to equations (2.5) and (2.7) satisfy the two Maxwell equations (2.1) and (2.3), respectively. The dynamic behavior of $\vec{A}$ and ψ is determined by the other two Maxwell equations (2.2) and (2.4), and these can be rewritten in terms of potential[1] as:

$$\text{div grad } \psi + \frac{\partial}{\partial t}(\text{div}\vec{A}) = -\frac{\rho}{\varepsilon} \tag{2.8}$$

---

[1] Applying vector identity: rot rot $\vec{A}$ = grad div $\vec{A}$ − ◊$\vec{A}$
   (◊ is the vector laplace operator [25][26])

$$\text{grad div}\vec{A} - \Diamond \vec{A} = \mu\vec{J} - \varepsilon\mu \, \text{grad}\frac{\partial \psi}{\partial t} - \varepsilon\mu\frac{\partial^2 \vec{A}}{\partial t^2}$$

$$\text{grad}\left(\text{div}\vec{A} + \varepsilon\mu\frac{\partial \psi}{\partial t}\right) = \mu\vec{J} + \Diamond\vec{A} - \varepsilon\mu\frac{\partial^2 \vec{A}}{\partial t^2} \, . \tag{2.9}$$

Applying the Lorenz condition:

$$\text{div}\vec{A} + \varepsilon\mu\frac{\partial \psi}{\partial t} = 0, \tag{2.10}$$

to equations (2.8) and (2.9) leads to two wave equations, one for ψ and the other for $\vec{A}$, respectively.

$$\text{div grad}\psi - \varepsilon\mu\frac{\partial^2 \psi}{\partial t^2} = -\frac{\rho}{\varepsilon} \tag{2.11}$$

$$\Diamond\vec{A} - \varepsilon\mu\frac{\partial^2 \vec{A}}{\partial t^2} = -\mu\vec{J} \tag{2.12}$$

2.1.2   Dosimetry and interaction mechanisms of ELF fields

Dosimetry is based on the relationship between dose (energy absorbed per unit mass) and biological effect, for ionizing radiation with sufficient energy to remove electrons from atoms [13]. Since the energy absorbed is directly related to the internal electromagnetic (EM) fields, i.e. electromagnetic fields within the object, dosimetry is interpreted as a means of determining EM fields within the biological body [14]. Dosimetry in this sense involves the measurement of the internal fields or determination through calculation of the internal fields, e.g. the induced current density in the low and intermediate frequency ranges (up to 100 kHz) and specific absorption rate (SAR) in the high frequency range (above 100 kHz). By general environmental exposure, the internal fields (rather than the incident fields) are responsible for interactions with biochemical systems independently of whether these interactions are thermal or non-thermal.

Depending on the size and shape of the object, its electrical properties, its orientation with respect to the incident fields and its operating frequency, the internal and incident EM fields can differ considerably. A biological effect occurs when exposure to EM fields causes some noticeable or detectable physiological alterations within a living system. Such effects may sometimes but not always lead to an adverse state of health which means that physiological changes have occurred which exceed the normal range for a brief period or more prolonged period of time. This negative health conditions develops when the biological effect is beyond that of the normal range so that the body is unable to compensate. Effects that are detrimental to health are often the result of biological effects that accumulate over time and depend on the degree of exposure. Therefore, detailed knowledge of the biological effects is important in determining the possible health risks generated by such exposures [14].

In daily life, we are exposed to extremely low frequency (ELF) fields from many sources, e.g. transmission lines, power tools and various electrical appliances. Usually the human body size is much shorter than the ELF wavelength (λ_0) [15] and also skin depth (δ) is much larger than human body size ($\lesssim 2$ m)

$$\text{Body size} \ll \delta.$$

Under such low-frequency conditions (quasi-static), the electric currents induced within the body tissues are much higher than the displacement currents, due to:

$$\sigma \gg \omega \, | \, \epsilon \, |,$$

and the electric and magnetic fields may be considered as independent components (where σ, ϵ and ω, respectively, represent the conductivity and permittivity of the tissue and angular frequency of the incident time harmonic fields). This differs from the situation regarding high (radio) frequency fields, where the electric and magnetic fields are inextricably linked ([16],[17],[18]). It can be shown from Faradays and Amperes laws that the secondary magnetic field produced in tissue by the current flow induced by the external magnetic field can be neglected, and, also that under quasi-static conditions, the permittivity values of the body tissues are non-significant. Furthermore, the biological tissues are non-magnetic, and, thus the permeability of the body tissues is equal to the free space value μ_0. Therefore, the body tissues can be regarded as conductive media with an ELF field that produces a current within the incident human body ([19],[43]). The magnitudes and spatial patterns of these induced currents depend on the incident magnetic field (i.e. frequency, magnitude and polarization) and also on the specific characteristics of the human body (i.e. size, shape and electrical properties of the tissues).

At a microscopic level [20] all tissues are composed of cells and extracellular fluids as shown in figure 2.1(a). The cell consists of two distinct parts the outer isolating membrane and the inner cytoplasm and nucleus. In the low frequency range, the impedance of the isolating membrane is higher than that of inner part. Therefore, the current induced in the tissues flows around the cell (in the extracellular fluid, which has a high conductivity), but not within the cell. On the other hand, in the high frequency range, the impedance of the isolating membrane is low: therefore, the current induced in the tissues flows both outside and within the cell (see figure 2.1 (b)). Such an induced current may cause a certain effect within the biological system in question ([21],[22],[23]).

It is possible to accurately calculate the induced current in a human body by applying numerical simulations (described in the next section). However, for simple geometric models (e.g. spherical, ellipsoid and two-dimensional disk models), which approximate the form of a human body, the induced current distribution in the body can be analytically calculated. For example, the current density in a two-dimensional

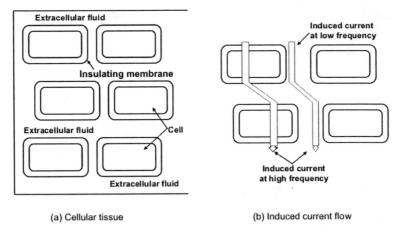

(a) Cellular tissue   (b) Induced current flow

Figure 2.1: Cellular tissue and induced current flow.

circular disk exposed to a uniform sinusoidal magnetic field (penetration perpendicular to the disk), can be easily determined using Faradays law of induction:

$$\oint_C \vec{E} \cdot \mathrm{d}\vec{l} = -\int_S \frac{\partial \vec{B}}{\partial t} \cdot \hat{n}\, \mathrm{d}s \qquad (2.13)$$

where $\vec{E}$ is the induced electric field vector, $\mathrm{d}\vec{l}$ is a vector of incremental length along the closed contour $\vec{l}$ enclosing surface s, $\hat{n}$ is a unit vector perpendicular to the surface element area $\mathrm{d}s$, and $\vec{B}$ is the magnetic flux density vector. If the uniform magnetic flux density $\vec{B}$ is perpendicular to the surface s, the induced electric field $\vec{E}$ along the closed circular disk of radius x is:

$$|\vec{E}| = \frac{\omega\, |\vec{B}|}{2}\, x \qquad (2.14)$$

where $\omega = 2\pi f$ is the angular frequency of the incident field. The current density strength $|\vec{J}|$ along the closed circular disk is then given by:

$$|\vec{J}| = \sigma\, |\vec{E}| = \sigma\, x\, f\, |\vec{B}|\, \pi \qquad (2.15)$$

where σ is the conductivity [24]. The secondary magnetic field produced by the current in the conductive medium is not taken into account. The error due to this simplification is sufficiently small, as long as the dimensions of the human body are small enough compared to the wavelength of the incident field. Furthermore, if the properties of the biological system remain constant, the induced current is directly proportional to the frequency of the applied field. However, the value of the induced current in the body based on the above equation is limited. This is because of the induced currents usually interface between the different layers in a heterogeneous object and are quite different from those predicted analytically for a homogeneous case.

2.1.3 ELF fields and cancer

Cancer is a term that is used to cover at least 200 different disorders, all of which involve uncontrolled cell growth. It is associated with uncontrolled mitosis, in which cells randomly divide and grow after escaping the body's normal control mechanisms. A primary disorder of cellular growth and differentiation, cancer can essentially be viewed as due to genetic malfunctioning at a cellular level. Thus the development of this disease is triggered within in the cell itself rather than in the body as a whole. The transformation of healthy cells into malignant ones is a complex process, which includes at least three distinct stages driven by a series of injuries to the genetic structure of the cells. This process can be described by a multi-step cancer development model [54], as shown in figure 2.2. The first step involves damage (injury) to the DNA cells, thereby causing the initial stage of cancer development. This is an irreversible phenomenon, in which certain agents are responsible for causing genetic mutations. The next step is that of the proliferation of already damaged cells. It is extremely unlikely that a single genetic injury to the cell will result in cancer; rather it appears that a series of genetic mutations are required. Damage inflicted to the DNA may affect various types of cells, and thus may cause more than one kind of cancer [14].

A large number of studies have reported that no significant damage to the DNA has been confirmed under most of ELF exposure conditions ([56]-[76]). These studies have shown that ELF fields do not cause DNA damage under the typical environmental exposure levels (i.e. below 100μT [82]). Furthermore, no DNA damage at exposure levels less than 1mT has been reported for the bacteria or yeast cells ([83],[84]). However, a few studies ([77]-[79]) have reported DNA damage in animal experimental system, but most of these have lacked the similar environmental exposure conditions, or have not even been replicated. Other studies have indicated that

ELF fields might also have some epigenetic activity ([55],[80],[81]). Possible evidence of these possible effects has been reported for magnetic field strengths above $100\mu T$ [82]. At this point, it should be noted that the main drawback in the majority of the experimental findings is that most of the studies have not been replicated, and the latter an essential aspect as it is the only means of verifying the initial results. As mentioned above, the energy associated with the environmental ELF fields is not high enough to cause direct damage to the DNA; however, indirect effects could be involved with ELF altering processes within the cells and which could lead to DNA breaks (see figure 2.3). Moreover, ELF fields well above environmental field intensities might enhance DNA synthesis, alter the molecular weight distribution during protein synthesis, delay the mitotic cell cycle, and induce chromosomal aberrations. One possible interaction hypothesis currently under investigation is that exposure to ELF fields could suppress the production of melatonin, located deep near the center of the brain. Melatonin is produced mainly at night and released into the blood-stream and so circulates throughout the body. It permeates into almost every cell in the human body, destroying free radicals and helping cell division to take place with undamaged DNA. Also it reduces the secretion of tumor-promoting hormones and protects the body from infection and the development of the cancer cells. It is therefore postulated that various cancers might proliferate if melatonin levels are lowered in the body. Several studies have reported reduced melatonin levels in the cells of animals and humans exposed to ELF fields. The effect varies according to the duration of exposure and the strength of the ELF fields. Formal cellular studies ([85]-[90]) have shown that in human breast cancer cell cultures, low level ELF magnetic fields (in the order of $1.2\mu T$) can block melatonin's ability to suppress the cancer cells [14].

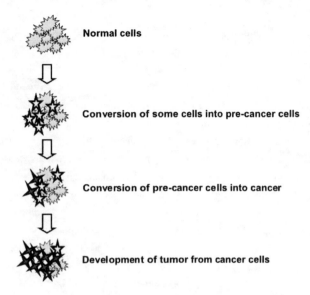

Figure 2.2: Conversion of certain cells into cancer cells.

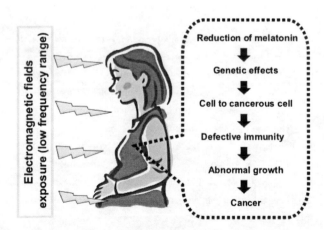

Figure 2.3: Stages in the development of cancer.

2.1.4   Protection guideline for electromagnetic fields

The International Commission on Non-Ionizing Radiation Protection (ICNIRP) [1]
has published the guidelines for limiting exposure to electric, magnetic and electromagnetic fields. It has recommended the frequency-dependent limits and also
included levels, from which a risk assessment for exposure to electric and magnetic
fields can be made. The limits are intended to restrict the degree of exposure, and
have been based on effects that have been established as detrimental to health, which
are termed basic restrictions. Depending on frequency, the physical quantities used
to specify the basic restrictions on exposure to low and intermediate frequency ranges
(up to 100 kHz) are the induced current densities. Ensuring protection against adverse health effects means that these basic restrictions have to be respected. The
reference levels of exposure are included for in the guideline provide a benchmark for
comparison with measured values of physical quantities; compliance with the former
will thereby ensure that the basic restrictions are kept to. However, if the measured
values are higher than those of the reference levels, it does not necessarily follow
that the basic restrictions have been exceeded, but it does indicate that a more
detailed analysis is necessary to assess compliance with these. Figure 2.4 shows the
basic restrictions (current density) and reference levels (magnetic flux density) for
the general public exposure to frequencies up to 100 kHz, which have been referred
to this study.

Furthermore, tables 2.1 and 2.2 respectively indicate the basic restrictions and the
reference levels for the exposure of the general public for the defined total frequency
range. It is sig-nificant to note that these guidelines are only designed to avoid the
immediate high-level risks; however, they are not applicable for prolonged low level
exposures, which may satisfy the re-ference level [14]. However, several institutions
have criticized these guidelines as lacking a clear interpretation of safety consid-
erations regarding exposure or direct application to equipment in use. A number
of concerns have been expressed over the inclusion and implementation of safety
factors, precautionary aspects, and the effect of long-term exposure. Accordingly, a
statement clarifying the manner in which the protection guidelines should be applied
in a regulatory and legislative context was presented by Matthes [48].

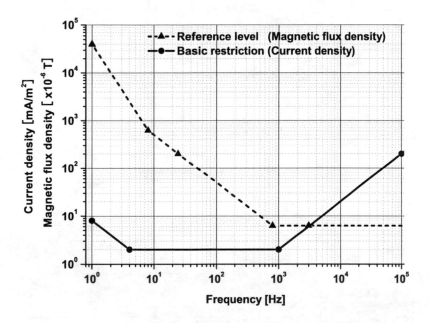

Figure 2.4: Basic restrictions and reference levels up to 100 kHz (for the general public).

Frequency Range	Current density (head and trunk) $(mA/m^2)(rms)$	Whole-body average SAR (W/Kg)	Localized SAR (head and trunk) (W/Kg)	Localized SAR (Limbs) (W/Kg)
up to 1 Hz	8	-	-	-
1-4 Hz	$8/f$	-	-	-
4 Hz-1 k Hz	2	-	-	-
1-100 kHz	$f/500$	-	-	-
100 kHz-10 MHz	$f/500$	0.08	2	4
10 MHz-10 GHz	$f/500$	0.08	2	4

Note: f is the frequency in [Hz]

Table 2.1: Basic restrictions for the general public.

Frequency Range	Magnetic field strength (A/m)	Magnetic flux density (μT)
up to 1 Hz	3.2×10^4	4×10^4
1-8 Hz	$3.2 \times 10^4/f^2$	$4 \times 10^4/f^2$
8-25 Hz	$4000/f$	$5000/f$
0.025-0.8 kHz	$4/f$	$5/f$
0.8-3 kHz	5	6.25
3-150 kHz	5	6.25
0.15-1 MHz	$0.73/f$	$0.92/f$
1-10 MHz	$0.73/f$	$0.92/f$
10-400 MHz	0.073	0.092
400-2000 MHz	$0.0037/f^{1/2}$	$0.0046/f^{1/2}$
2-300 GHz	0.16	0.20

Note: f as indicated in the frequency range column

Table 2.2: Reference levels for the general public.

2.2   Theoretical basis

2.2.1   Equivalent source model

It is necessary to describe the magnetic field around the electrical appliances accurately in order to evaluate possible hazards due to exposure by numerical means. However, the full geometry-based numerical modeling of a real magnetic source (e.g. the motor of the electronic appliance) is extremely complex. One method to address this issue is the use of equivalent sources. Efficiency can be achieved by using simplified numerical models that include a limited number of simple sources (in this study, the magnetic dipole moments) and adequately representing the electromagnetic fields of the real magnetic source[2].

The equivalence principle:
Figure 2.5 shows a simple application of the equivalence principle. Suppose that we have a set of sources (e.g. motor, eddy coil and antenna) which is producing either the electric field $\vec{E}$ or the magnetic field $\vec{H}$ throughout the free space. In this free space, an arbitrary closed surface S is defined such that only free space exists external to S, as shown in figure 2.5 (a). A second problem equivalent to the original problem external to S is then set up as follows. We specify the field external to S to be the same as in the original problem. We specify the field internal to S to be zero. Both choices are possible solutions to the source-free electric and magnetic field equations in the free space (the external field is the field of the original problem; the internal field is the zero field). The surface currents $\vec{J}_*$ and $\vec{M}_*$ must exist on S in order to support such a field. The surface currents are expressed as follows:

$$\vec{J}_* = \hat{n} \times \vec{H} \qquad \vec{M}_* = -\hat{n} \times \vec{E}, \tag{2.16}$$

where $\hat{n}$ points outwards from the surface S, and $\vec{E}$ and $\vec{H}$ are those from the original problem ([27],[28]).

In this study, these equivalent electric surface currents $(\vec{J}_*)$ are replaced with the magnetic dipole moments, which compose the proposed source model (i.e. equivalent

---

[2] **Huygens' Principle:** Each point on a primary wavefront can be considered a new source of a secondary spherical wave and a secondary wavefront can be constructed as the envelope of these secondary spherical waves. With electromagnetic fields, the electric and magnetic surface currents can be assumed to act as sources of the secondary waves [26].

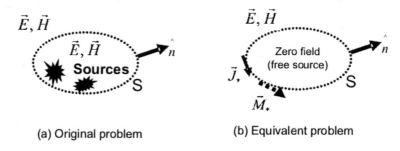

(a) Original problem   (b) Equivalent problem

Figure 2.5: The equivalence principle and the surface currents.

source model) described in the subsequent sections. The following equation represents the relation between surface current and magnetic dipole moment ([29],[30]):

$$\vec{m}(i) \;=\; \mu_0 \, \vec{J}_*(i) \, \mathrm{d}s, \qquad\qquad (2.17)$$

where (i) is the identification number of discrete element area $\mathrm{d}s$ on the surface
S, $\vec{m}(i)$ is the $(i)^{th}$ magnetic dipole moment, and $\vec{J}_*(i)$ is the equivalent surface
electric current on the $(i)^{th}$ surface element $\mathrm{d}s$. In the theory of the equivalent
source model, the magnetic dipole moments (instead of equivalent surface currents)
are numerically calculated from the measured magnetic field data.

Basics of the equivalent source model:

As shown in figure 2.6, the equivalent source model is applied in three stages. In the first step, the magnitude and the phase of the components of the magnetic field vectors are determined on an arbitrary test surface[3] around the device under test. Due to the extreme large wave length in the low and intermediate frequency ranges, the relative distances in a close region around the appliance are usually not significant, which enables one to neglect the phase of the field. However, to be able to calculate the circular field polarization emitted by the appliance (what occurs rarely), the phase of the magnetic field vectors must be measured and considered in the numerical calculation in the following step. In the second step, a numerical field transformation is carried out, to obtain the dipole moments of an equivalent distribution of magnetic elementary dipoles according to the magnetic field data obtained in the first step. The magnetic field $\vec{H}_d$ of a magnetic elementary dipole is given by equation (2.18) ([29],[30]). Consequently, the magnetic field $\vec{H}_{all}$ resulting from a number of magnetic dipoles N with corresponding moments $\vec{m}_i$ is given by equation (2.19). In our study, $\vec{H}_{all}$ is the given magnetic field data on the test surface, while $\vec{m}_i$ represent the unknown magnetic dipole moments to be determined numerically (one approach is to use the conjugate gradient method ([50],[51],[52])). The accuracy of the equivalent source model increases with the number of given magnetic field data points, while the numerical determination of the unknown magnetic dipoles requires that the given magnetic field data points must be equal or exceed the number of unknown magnetic dipoles. As shown in equations (2.20) and (2.21), both $\vec{H}_{all}$ and $\vec{m}_i$ have a phase term which allows the correct treatment of polarization.

$$\vec{H}_d(\vec{r}) = -\text{grad}\left(\frac{\vec{m} \cdot (\vec{r} - \vec{r}_0)}{4\pi\mu_0|\vec{r} - \vec{r}_0|^3}\right) \tag{2.18}$$

$$\vec{H}_{all}(\vec{r}) = \sum_{i=1}^{N}\left\{-\text{grad}\left(\frac{\vec{m}_i \cdot (\vec{r} - \vec{r}_{0,i})}{4\pi\mu_0|\vec{r} - \vec{r}_{0,i}|^3}\right)\right\} \tag{2.19}$$

$$\vec{H}_{all} = \vec{H}_{all} \cdot \exp(\,\mathrm{j}\varphi_{H_{all}}) \tag{2.20}$$

$$\vec{m}_i = \vec{m}_i \cdot \exp(\,\mathrm{j}\varphi_{m_i}) \tag{2.21}$$

---

[3]For practical reasons, often the surface of a cylinder is used.

In these equations, $\vec{r}$ is the observation point, while $\vec{r}_0$, $\vec{r}_{0,i}$ represent the positions of the magnetic dipole moments; $\varphi_{H_{all}}$, φ_{m_i} are the phases of the magnetic field and magnetic dipole moment, respectively. Owing to the peculiarities of the numerical approach (equation 2.19), it is not possible to locate the magnetic dipole moments at the positions where the magnetic fields are defined. Therefore, the magnetic dipole moments are e.g. defined on the surface of a cylinder with a smaller diameter. The reduced diameter of the equivalent source model is less than 80% of the diameter of the first cylinder, from which the magnetic field data were collected in the first step [31]. Furthermore, the convergence of the total magnetic dipole numbers is confirmed with the relative difference value of the maximum magnetic fields on the cylinder surface, which is obtained from the measured and the calculated (i.e. equivalent source model) results. In this study, the maximum difference value is set to be less than ±3%.

Finally, in the third step, the equivalent numerical model is applied in a numerical calculation tool to investigate the magnetic field property or dosimetry of human exposure. In our study, we use the simulator FEKO ([32],[33]) for the field calculation and the scalar potential finite difference (SPFD) method to determine the current density induced in the human body models, which is described in the next section.

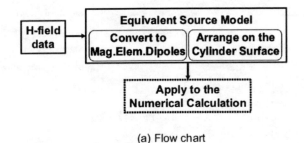

(a) Flow chart

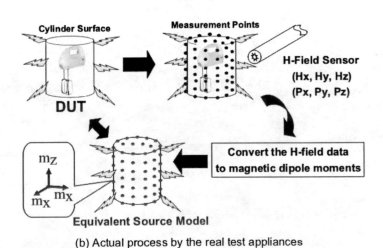

(b) Actual process by the real test appliances

Figure 2.6: General process of calculation procedure.

2.2.2 Numerical method for induced current calculation

In this study, the scalar potential finite difference (SPFD) method ([42]-[45]) is used for dosimetric calculations. This method is effective for calculating the induced current density in the numerical conductive human body model, whose dimensions are much smaller than the wavelength of the incident fields and the internal skin depths (up to 100 kHz). Furthermore, the secondary magnetic field and other higher-order terms in the body model, which are produced by the induced current density can be neglected under the low frequency assumptions as described in section 2.1.2. In the case of use of this method in the higher frequency range, the skin depth becomes comparable with the body dimensions and the phase of the magnetic field must be taken into account([46],[47]).

In the fundamental theory of the SPFD method, two types of sources contribute to the internal electric field (inside the body model). One is the surface charge distribution, which occurs with the external electric field (outside the body), and the other is the external magnetic field, which induces eddy currents in the body model. In this study, the magnetic field produced by appliances (i.e. magnetic source: motor, eddy coil) is significant, while the electric field can generally be neglected. Consequently, the contribution of the external electric field is ignored in this study, and we consider only the internal electric field $\vec{E}$ and the associated current density $(\vec{J})$ induced by the external magnetic field $\vec{B}$, which is generated by the magnetic source [42]. Then:

$$\text{rot } \vec{E} = -j\,\omega\,\vec{B} \tag{2.22}$$

$$\text{div}\left(\sigma\vec{E}\right) = 0. \tag{2.23}$$

The magnetic fields from the source (e.g. elementary dipole moments or coil model) can be expressed in terms of a magnetic potential $\vec{A}$, and hence the equation (2.5) transforms into:

$$\text{rot}\left(\vec{E} + j\,\omega\,\vec{A}\right) = 0, \tag{2.24}$$

which implies that:

$$\vec{E} = -\text{grad}\,\psi - j\,\omega\,\vec{A} \tag{2.25}$$

where ψ is a scalar potential. Substituting $\vec{E}$ into equation (2.23), yields the differential equation:

$$\text{div}\left(\sigma \operatorname{grad} \psi\right) = \text{div}\left(-\mathrm{j}\,\omega\,\sigma\,\vec{A}\right). \tag{2.26}$$

As shown in the above equations, the applied magnetic field source is incorporated as a vector potential term in the electric field. Due to the quasi-static conditions described above, the phase of the vector potential $\vec{A}$ and the electric field $\vec{E}$ are not considered in this study.

In the numerical implementation, the potential ψ is considered to be defined at the cell nodes, and electric field strength components (E_x, E_y, E_z) are defined parallel (i.e. centered) to the cell edges (see figure 2.7). A finite difference approximation for equation (2.26) at a given node can then be constructed by applying the divergence theorem to a hypothetical shifted cell with the given node at the center of cell. As shown in figure 2.8, a local indexing scheme is applied, where the target node is labeled as zero (i, j, k) and both the nodes and edges connected to it on the $+x$, $-x$, $+y$, $-y$, $+z$ and $-z$ sides are indexed from 1 (i+1, j, k) to 6 (i, j, k-1), respectively. The derivation of the finite difference equation is given by:

$$\psi_0 = \left(\sum_{n=1}^{6} s_n \left(\psi_n - (-1)^n \, l_n \, j\omega A_n\right)\right)\left(\sum_{n=1}^{6} s_n\right)^{-1} \tag{2.27}$$

where,

$n=1$	$s_n = \sigma_x^{i+1}\,l_x$	$(-1)^n\,l_n A_n = -l_x A_x^{i+1}$	$\psi_n = \psi^{i+1}$
$n=2$	$s_n = \sigma_x^{i-1}\,l_x$	$(-1)^n\,l_n A_n = +l_x A_x^{i-1}$	$\psi_n = \psi^{i-1}$
$n=3$	$s_n = \sigma_y^{j+1}\,l_y$	$(-1)^n\,l_n A_n = -l_y A_y^{j+1}$	$\psi_n = \psi^{j+1}$
$n=4$	$s_n = \sigma_y^{j-1}\,l_y$	$(-1)^n\,l_n A_n = +l_y A_y^{j-1}$	$\psi_n = \psi^{j-1}$
$n=5$	$s_n = \sigma_z^{k+1}\,l_z$	$(-1)^n\,l_n A_n = -l_z A_z^{k+1}$	$\psi_n = \psi^{k+1}$
$n=6$	$s_n = \sigma_z^{k-1}\,l_z$	$(-1)^n\,l_n A_n = +l_z A_z^{k-1}$	$\psi_n = \psi^{k-1}$

l_x, l_y and l_z denote the edge lengths of the cubic cell and i, j, k are the node indices in the x, y, and z directions. The potentials ψ_n at the cubic cells (constructing the numerical human body model) are solved using an iterative -the successive over-relaxation (SOR) technique ([50],[51],[52]). By this iterative calculation, a zero value is applied to the initial condition. Finally, the electric fields are obtained from equation (2.25), which is used for calculating the induced current density $(\vec{J} = \sigma \, \vec{E})$ inside the numerical human body model.

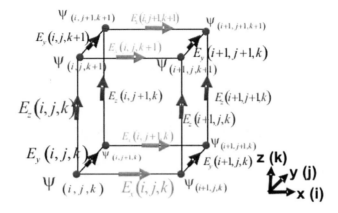

Figure 2.7: Definition of the potential and E-fields positions on the cell.

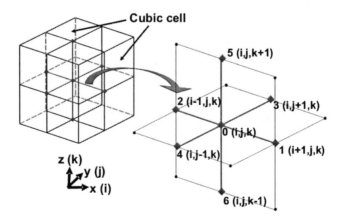

Figure 2.8: Local indexing scheme at a node.

Chapter 3

Equivalent source model and magnetic field

3.1 Validation of the numerical calculation

3.1.1 Example of the numerical model

This section defines the sample used to investigate the accuracy of the proposed source model and the calculation procedure. As shown in figure 3.1, a numerical model is considered for this purpose. Three circular current loops, orthogonally arranged with identical center point, are used to simulate the magnetic field characteristics similar to those of an electric appliance. Loop diameter is 0.04 m and a current of 1 A is impressed on the loops at a frequency of 50 Hz. The values of the magnetic field are collected on the surface of a cylinder with diameter $D = 0.06$ m and height $L = 0.24$ m (± 0.12 m). For the magnetic field calculations, we use the simulator FEKO [32]. As outlined in Section 2.2.1, the magnetic dipoles (total number of 214 dipoles) of the equivalent source model are located on the surface of a cylinder.

3.1.2   Convergence of the source model

The measurement system (see figure 3.6), which is described later in section 3.2.1 ignores, the cover plates (bottom and top area) and the cylindrical surface is used to collect the field values. Therefore, we have carried out an investigation to study the impact of this omission; numerical simulations are carried out using various values of height L and the resulting error is assessed. Figure 3.2 shows the relative magnetic field error (defined as the difference between the reference field of the current loop and the output achieved by the equivalent source model determined without the consideration of cover plates) versus the relative height L/D of the cylinder, i.e. the length normalized by diameter (D = 0.06 m). The observation point for the magnetic field is placed at the surface of the cylinder, where the maximum magnetic field appears (Theta=90 deg). As expected, the error decreases as (L/D) increases; the error is less than 1% for L/D = 3...4. This result indicates that the magnetic fields on the bottom cover can be ignored without a significant loss of accuracy for L/D values of 4 or more. Figures 3.3 (a) and (b) show the distribution of the magnetic flux density ($|B| = \mu_0|H|$) on the cylinder surface directly calculated for a current loop (as the reference) and the equivalent source model, respectively. The cylinder's length is four times the diameter (D = 0.06 m), i.e. L = 0.24 m. As shown in these field patterns, two hot spots appear at 90 deg and 270 deg around H = 0.0 m. Even for this relatively complicated field pattern, the distribution and the maximum $|B|$ value strength show good agreement. Hence, it is shown that the equivalent source model accurately reproduces even complicated field patterns around a magnetic field source.

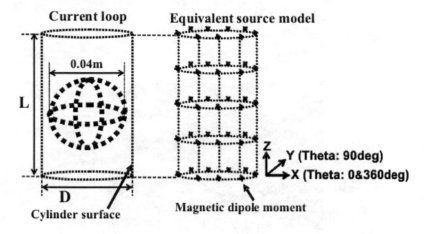

Figure 3.1: Current loop and equivalent source model.

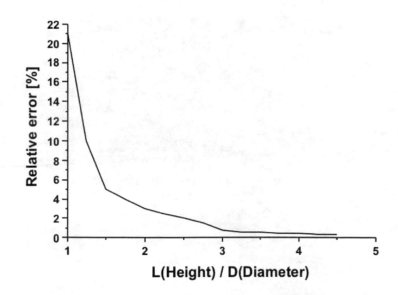

Figure 3.2: Magnetic field error versus relative cylinder height.

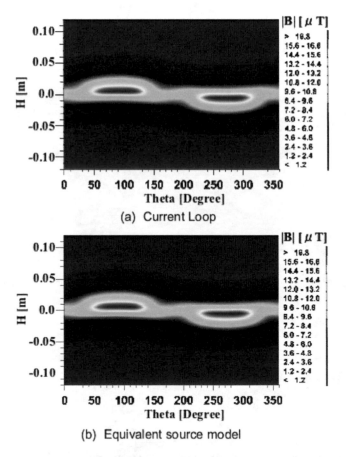

Figure 3.3: Magnetic flux density distribution on the cylinder surface (a: Current loop, b: Equivalent source model).

3.1.3 Magnetic field distribution and field vectors

Figures 3.4 (a) and (b) show the magnetic field strength |H| [dB A/m] distribution on a square area of 0.4 m x 0.4 m (YZ-plane at X = 0.0 m and XY-plane at Z = 0.0 m) in front of the current loop, respectively. In this result, the maximum magnetic field point (Theta = 90 deg) on the cylinder surface, which is shown in figure 3.3 arranged on the Y axis. In this result, the main beam (direction of the field intensity) lies on a line that is approximately the oblique angle of 60 degrees above the positive Y axis. Furthermore, the maximum difference in magnetic field strength is less than 1% for the whole calculated area. Figure 3.5 shows the magnetic field strength vectors (Hx, Hy, Hz) along the Y axis for the current loop and for the equivalent source model. As shown in this picture, the main magnetic field strength vectors are Hy and Hz (H-polarization lies in the YZ-plane) and the field strength vectors again show a good agreement between the two models. This good agreement validates the equivalent source model. Furthermore, it is confirmed that the equivalent source model reproduces almost identical (less than 1% error) the main beam, the field strength, and also the field strength vectors. This would be difficult to achieve if only a simple source model was applied.

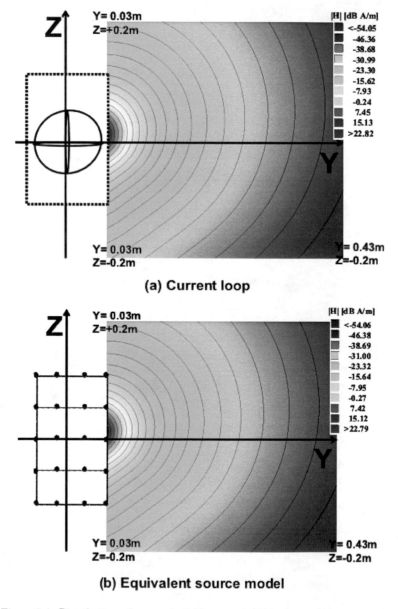

Figure 3.4: Distribution of magnetic field strength |H| in front of the current loop (a:YZ-plane at X = 0.0 m), b:XY-plane at Z = 0.0 m).

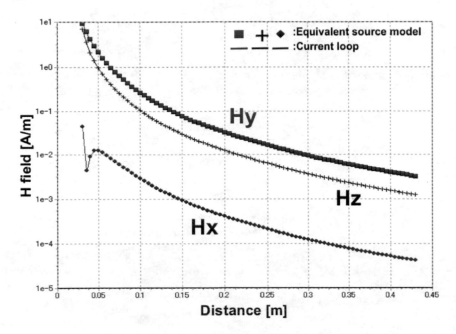

Figure 3.5: Magnetic field vectors along the Y axis (X = Z = 0.0 m).

3.2   Validation using real appliances

In the previous section, the validity of the equivalent source model is confirmed for a numerical source model. However, it is significant to validate it with real appliances for its application. In this section, the validity of the equivalent source model is verified for commercial appliances by comparing the results of the measured and calculated magnetic flux density around the devices.

3.2.1   Measurement system and sensor

The measurement of the magnitude and phase of the magnetic field around real electrical appliances is carried out using 3D-scan measurement system [34], which is shown in figure 3.6 (a). The 3D-scan mainly consists of a robotic arm with a magnetic field sensor mounted on the top and a personal computer controlling its position. Furthermore, a loop coil is arranged around the turntable for the measurement of the reference phase. The location of the measured magnetic field is determined by using the system on the closed surface of the cylinder enveloping the whole appliance. As shown in figure 3.6 (b), the magnetic field sensor is connected at the top of the robotic arm and consists of three orthogonal loops with a diameter of 2.0 cm. The magnetic field is measured conducted in the time domain and transformed to the frequency domain by using a DFT converter.

3.2.2   Test method

As shown in figure 3.7, an observation line is set vertical to the equivalent source model (cylinder form: height, diameter) surface, and the magnetic field distribution is measured and calculated along this line for the equivalent source model and real appliances, respectively. We used a commercial hair dryer (height = 72 cm, diameter = 10 cm) and a hand mixer (height = 77 cm, diameter = 15 cm) as real appliances.

Due to the large inhomogenity of the magnetic field close to the appliance, the magnetic field inside the loop area of the sensor is not uniform. Therefore, the calculated magnetic field strength vectors are averaged over a circular area, which has the same area (3 cm^2) as the measured sensor loop (see figure 3.6 (b)).

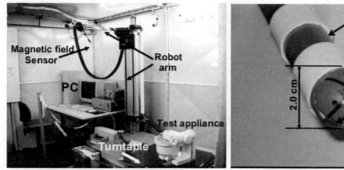

(a) Measurement system

(b) Magnetic field sensor
(coil size: 3 cm²)

Figure 3.6: Photograph of the 3D-scan measurement system and the isotropic magnetic field sensor.

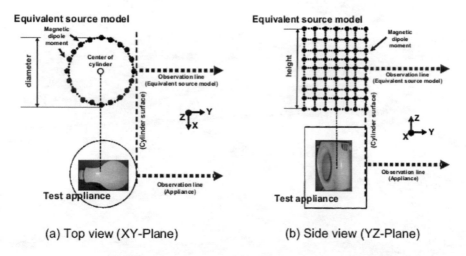

(a) Top view (XY-Plane)

(b) Side view (YZ-Plane)

Figure 3.7: Arrangement of the equivalent source model and test appliance.

3.2.3   Results

Figures 3.8 and 3.9 show the measured and calculated (i.e. the equivalent source model) results for the hair dryer and hand mixer, respectively. The X and Y axes show the distance from the cylinder surface and the magnetic flux densities |B|, respectively. The measured and calculated results are in good agreement, and the validity of the equivalent source model is confirmed for real household appliances. However, the maximum difference (~15%) between both results is observed for small distances around 1 cm. This difference is caused by the inaccurate arrangement of the sensor top by the measurement system, which induces the gap of observation position (±1 cm) between the measured and calculated results. However, the reduction in the difference is observed by replicated measurements, which are conducted for close distances (1 to 3 cm) from the cylinder surface by rearranging the sensor position manually. Due to these measured results, the maximum difference decreases to 5 to 10% of the range.

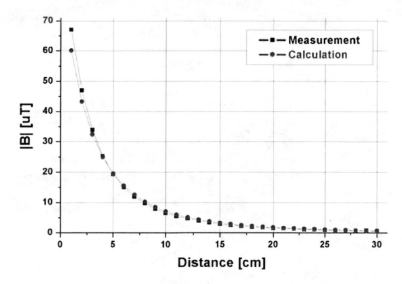

Figure 3.8: Magnetic flux density against distance (hair dryer).

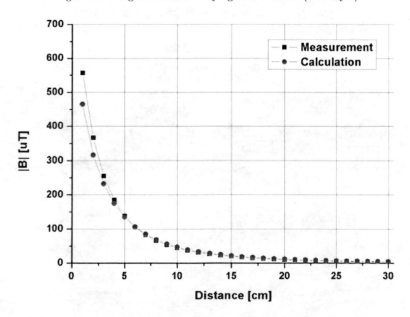

Figure 3.9: Magnetic flux density against distance (hand mixer).

3.3   Magnetic field properties around the appliances

In this section, the magnetic field properties at low frequency (50 Hz) and intermediate frequency (21 kHz) for the test appliances described in section 3.3.1 are investigated using the coil model, which is prescribed as a substitute source model in the European standard EN50366 [5]. The accuracy of the magnetic field strength vectors and the distribution obtained using the coil model are compared with the results of a more realistic model of the test appliances obtained from the equivalent source model.

3.3.1   Test appliances and magnetic field measurement

Measurement at low frequency of 50 Hz

Figure 3.10 shows some commercial appliances, whose main frequency of leakage magnetic field is 50 Hz and which are used as test appliances in this study. The equivalent source model is calculated for each appliance with the same process as described in section 2.2.1. The parameters of the equivalent source model, such as the diameter, height and the number of dipoles for each appliance, are given in table 3.1.

(a) Hand mixer   (b) Drilling machine

Figure 3.10: Overview of the test appliances.

	Diameter [m]	Height [m]	Number of dipole moment
(a) Hand mixer	0.22	0.5	1007
(b) Drilling machine	0.2	0.45	2338

Table 3.1: Parameters for the equivalent source model.

Figure 3.11 (a) shows the measured magnetic flux density ($|B|$) distribution of the hand mixer on the cylinder surface. This is measured by using a 3D-scan measurement system for calculating the equivalent source model. Figure 3.11 (b) shows the magnetic flux density distribution on the cylinder surface, which is calculated from the equivalent source model. As shown in both results, the field strength diminishes to less than 10% of the maximum magnetic flux density ($|B|_{max}$) at the source height of ± 25 cm, and the $|B|_{max}$ of both results shows good agreement with the difference that is below 1%. As shown in the measured distribution, $|B|_{max}$ is 355 μT (appears in the direction of Theta = 90 deg), which exceeds the reference level (ICNIRP limit for general public: 100 μT at 50 Hz).

Figure 3.12 shows the $|B|$ distribution of the drilling machine similar to the hand mixer. As shown in this result, $|B|_{max}$ of both results are in good agreement with the difference below 1%. Furthermore, $|B|_{max}$ is 17.9 μT (under the reference level) and satisfies the ICNIRP guideline [1].

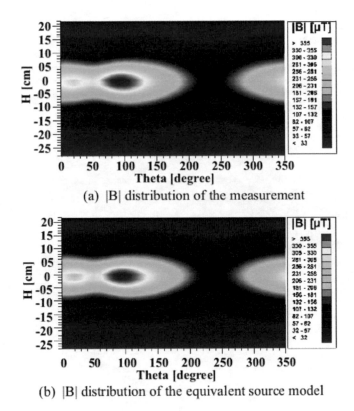

(a) |B| distribution of the measurement

(b) |B| distribution of the equivalent source model

Figure 3.11: Magnetic flux density distribution of the hand mixer.

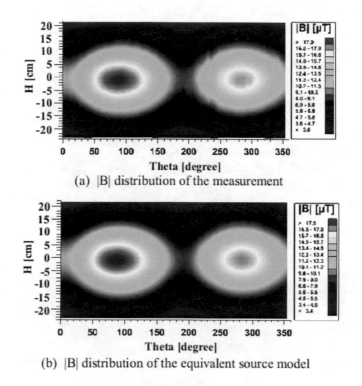

(a) |B| distribution of the measurement

(b) |B| distribution of the equivalent source model

Figure 3.12: Magnetic flux density distribution of the drilling machine.

Measurement at intermediate frequency of 21 kHz

As shown in figure 3.13, a commercial induction heater is selected as another test appliance, whose main frequency of the leakage magnetic field is 21 kHz. The magnetic field measurement for the equivalent source model is conducted as described before. In order to maintain constant output power of the induction heater (IH), the power is controlled by an inverter circuit. Consequently, the frequency of the peak in the magnetic field spectrum (leakage field around the appliance) changes during the operation. Hence, the measurement frequency range is considered to be ±5% (between 20 kHz and 22 kHz) from the center frequency of 21 kHz. Furthermore, the rise of the induction heater body temperature (the heat of boiled water in the vessel diffuses to the induction heater) deranges the normal operation of the inverter circuit, which results an excess of the peak spectrum over the ±5% frequency range. Therefore, the induction heater is allowed to cool down in a 5 minute interval to avoid a large variation in the frequency.

Figure 3.14 (a) shows the measured magnetic flux density |B| distribution on the cylinder surface (height = 80 cm, diameter = 54 cm). Theta = 0 [degree] and 90 [degree] indicate the frontal and side direction of the induction heater, as shown in figure 3.13. The measured peak frequency range is between 20.068 kHz and 21.694 kHz (difference: 1.62 kHz) and the difference from the center frequency of 21 kHz is approximately ±4%. As shown in this figure, the peak value appeared at the side of the appliance in the direction of theta equal to 90 deg, which is shown in figure 3.16. Figure 3.14 (b) shows the |B| distribution on the cylinder surface, calculated using the equivalent source model (height: 80 cm, diameter: 50 cm, number of dipoles: 4144) at the center frequency of 21 kHz. Comparing the results between the measured (a) and equivalent source model (b), a good agreement is observed for both maximum |B| value and the |B| distribution, although the measured frequency varies ±4% from the center frequency. Due to this good agreement, the validity of the equivalent source model is reconfirmed for the induction heater.

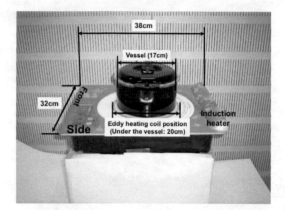

Figure 3.13: Overview of the induction heater.

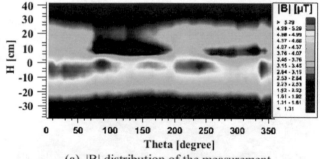

(a) |B| distribution of the measurement

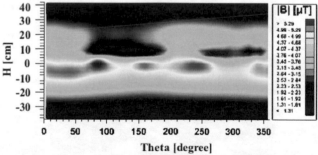

(b) |B| distribution of the equivalent source model

Figure 3.14: Magnetic flux density distribution of the induction heater.

3.3.2   Determination process of the coil radius

As described in section 1, a new standard measurement method (EN50366 [5]) for household appliances is currently enforced by the European committee for electrotechnical standardization (CENELEC). In this new standard EN50366, two evaluation methods are provided for household appliances, whose magnetic field levels exceed the limit of the reference level given by the ICNIRP guideline [1]. Hence it is necessary to assess compliance with the basic restrictions [1]. In the first evaluation method of EN50366, the generator (e.g. the motor or the eddy heating coil inside the test appliance) that creates the magnetic field around the appliance is approximated by the coil model. The radius of the coil model is determined so that the field strength caused by the coil model and the test appliance coincide. In order to obtain nearly the same field strength distribution using the approximated coil model, the optimum coil size (radius) is determined based on the measured magnetic field around the test appliance. Furthermore, the coil aspect is brought into accord with the one single axis direction of the main magnetic field intensity, which occurs in the same target direction (e.g. the human body model). The details of the determination procedure of the coil radius are described below.

In order to determine the coil radius (r_{coil}), the maximum magnetic field point (hot spot: $|B|_{max}$) on the surface of the induction heater (measurement surface is defined in EN50366) is determined in the first step, using a small magnetic sensor (as described in section 3.2.1) at the frequency of interest. After finding the hot spot, the magnetic field value is measured along a line (in the tangential plane) with an interval distance dr (0.01 m) starting (r = 0 m) from the hot spot until 10% of the maximum magnetic flux density value (r = X[m]: 0.1 $|B|_{max}$) is reached. The direction of the lowest gradient of the field is considered in this measurement to determine the wide exposure range using the coil model, which considers the worst case. In the second step, the measured magnetic flux density values are normalized and integrated in order to obtain the single value of G [m], as shown in the following equation (3.1)

$$G = \int_{r=0}^{r=X\{0.1\,|B|_{\max}\}} \frac{|B|(r)}{|B|_{\max}(r=0)}\, dr. \tag{3.1}$$

The radius of the coil (r_{coil}) is defined using this G value and the d_{coil} (see figures 3.15 and 3.16), which is the distance between the hot spot point on the surface of the appliance and the magnetic field generator (e.g. the motor inside the test appliance or the eddy heating coil inside the induction heater). A suitable coil radius is chosen from the table, which is indicated in EN50366. However, in the case of the induction heater, the magnetic field radiates from the surface area of the eddy heating coil (spiral coil), which has a diameter of 20 cm (see figure 3.13). Therefore, it is not possible to define the distance (d_{coil}) between the edge of the eddy heating coil and the surface of the appliance. Hence both the distance (d_{coil}) and radius (r_{coil}) of the coil are chosen directly from the table in EN50366 corresponding to the G value.

3.3.3   Alignment position of the source models

Figures 3.15 and 3.16 show the arrangement of the coil model and the equivalent source model, which are used for magnetic field calculation. The direction and height of the observation line are chosen from the hot spot on the test appliance, and the starting point (d = 0 m) is chosen from the surface of the equivalent source model to enable the comparison of the results of either source model. The parameters obtained by the advanced measurement process described in the previous section are listed in table 3.2. In this table, r_{coil} indicates the radius of the coil, d_{coil} is the distance between the surface of the test appliance and the magnetic field generator, $|B|_{\max}$ ($|H|_{\max}$) is the maximum magnetic flux density (magnetic field strength) and d_{cyl} is the distance between the test appliance and the cylinder of the equivalent source model surface as shown in figures 3.15 and 3.16. For the induction heater (c), two different coil models are investigated, which are selected from the description of EN50366 with the G value (0.0112 m); one is the smallest radius and the other is the largest radius. The parameters (diameter, height and the number of dipole moments) of the equivalent source model for each test appliances are described in the previous section 3.3.1. The magnetic field values around the coil model and the equivalent source model are independent of the frequency variation [30]. Hence, the single frequency of 21 kHz is conducted as main frequency for the induction heater and the actual frequency variation of ±4% is not considered in the magnetic field calculation. Furthermore, the impressed current (21 kHz) of the coil model is coordinated with the measured hot spot value at the distance of d_{coil}. A software tool FEKO [32] is used for the magnetic field calculations.

	G [m]	r_{coil}[m]	d_{coil}[m]	d_{cyl}[m]	$|H|_{max}$[A/m] ($|B|_{max}$[μT])
(a) Hand mixer	0.0426	0.02	0.05	0.055	282.5 (355.0)
(b) Drilling machine	0.0568	0.01	0.0725	0.14	14.24 (17.9)
(c) Induction heater					
(small coil)	0.0112	0.01	0.146	0.11	17.33 (21.8)
(large coil)	0.0112	0.07	0.109	0.11	17.33 (21.8)

Table 3.2: Parameters for the coil source model for appliances.

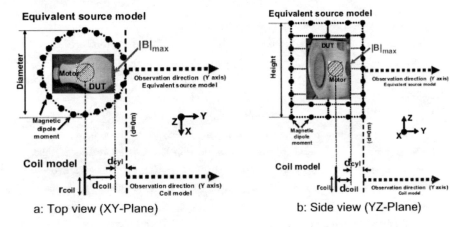

a: Top view (XY-Plane)   b: Side view (YZ-Plane)

Figure 3.15: Arrangement of source models applied in the magnetic field calculation (hand mixer and drilling machine).

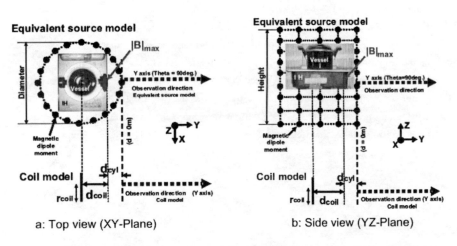

a: Top view (XY-Plane)   b: Side view (YZ-Plane)

Figure 3.16: Arrangement of source models applied in the magnetic field calculation (induction heater).

3.3.4   Results at the low frequency of 50 Hz

Figures 3.17 and 3.18 show the magnetic field strength |H| [dB A/m] pattern on the square area of 0.4 m × 0.4 m (XY-plane at Z = 0.0 m) in front of the coil model (a) and equivalent source model (b) for the hand mixer and drilling machine, respectively. As shown in these field patterns of the coil model (a), the field is symmetric along the Y axis and the direction of the field intensity is along the Y direction as prescribed. On the other hand, the field pattern of the equivalent source model (b) differs from the coil model. Particularly for the hand mixer (figure 3.17), the direction of the field intensity lies on a line that is approximately the oblique angle of 80 degrees above the negative X axis. This could be caused by the rotation of the motor inside the appliance. However, it is difficult to simulate these complicated field patterns using the coil model of EN50366 that defines the field intensity direction only for a single axis (Y). Figure 3.19 and 3.20 show the magnetic field strength vectors (Hx, Hy, Hz) along the Y axis (in figures 3.17 and 3.18). As shown by these results, the dominant field strength vector is Hy for all test appliances and a good agreement between the coil model and the equivalent source model is observed for Hy, with a maximum difference of +5 dB (+: coil value is larger) and +5.4 dB for drilling machine and hand mixer, respectively. Furthermore, the value of dominant field strength vector Hy of the coil model value is almost equal to that of field strength (|H|) value, due to peculiarly small values of the non-dominant (recessive) field strength vectors (Hx and Hz). On the other hand, the recessive field strength vectors of the equivalent source model do not significantly differ from the dominant vector and lead to a maximum difference of 2 dB (hand mixer) caused by the different field intensity direction. Hence, the coil model approximates the dominant field strength vector and the field strength with a good accuracy, but not the recessive field strength vectors.

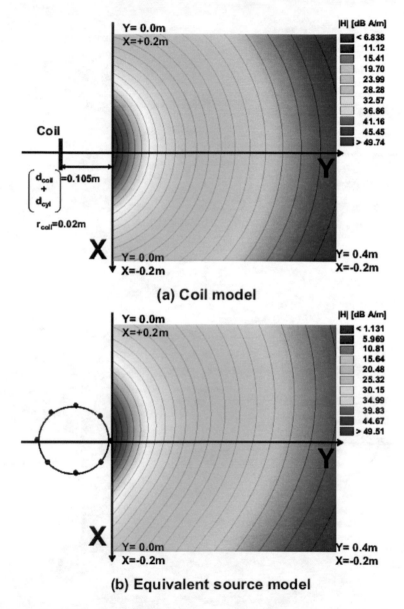

Figure 3.17: Distribution of magnetic field strength |H| in front of the (a) coil and (b) equivalent source model (hand mixer).

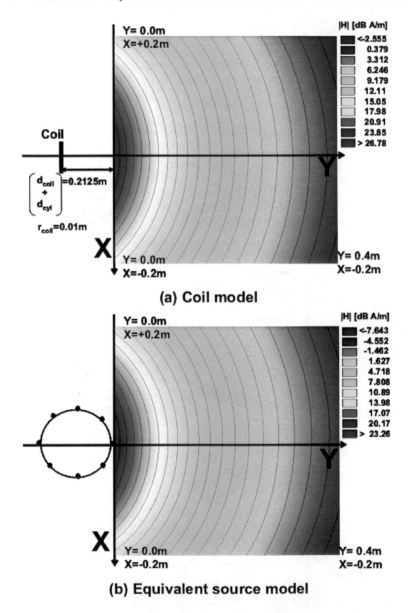

Figure 3.18: Distribution of magnetic field strength |H| in front of the (a) coil and (b) equivalent source model (drilling machine).

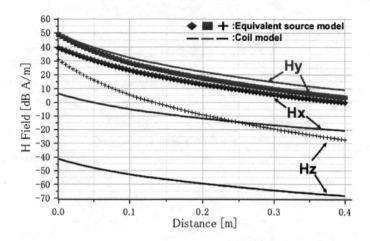

Figure 3.19: Magnetic field strength vectors along the Y axis at X = Z = 0.0 m (hand mixer).

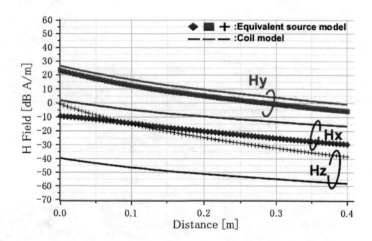

Figure 3.20: Magnetic field strength vectors along the Y axis at X = Z = 0.0 m (drilling machine).

3.3.5   Results at the intermediate frequency of 21 kHz

Figure 3.21 shows the magnetic field strength $|H|$ [dB A/m] pattern on a square area of 1.6 m × 1.6 m (YZ-plane at X = 0.0 m) in front of the coil model (a: $r_{coil} = 0.01$ m, b: $r_{coil} = 0.07$ m) and equivalent source model (c), respectively. As shown in the field patterns of both coil models, the field is symmetrical to the Y axis and the direction of the field intensity is in the Y direction as prescribed. On the other hand, the field pattern of the equivalent source model (c) is symmetrical with respect to the Y axis and two hot spots are observed nearby, which appear at the height of the induction heater body and the vessel, respectively.

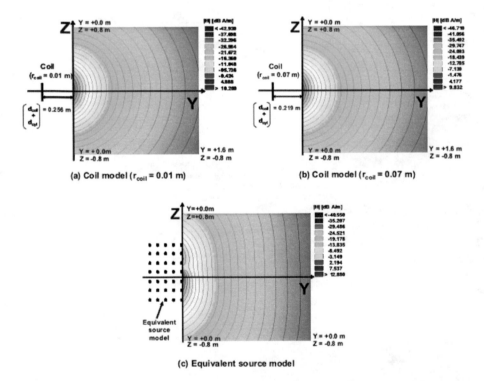

Figure 3.21: Distribution of magnetic field strength $|H|$ in front of the coil model (a: $r_{coil} = 0.01$ m, b: $r_{coil} = 0.07$ m and c: equivalent source model).

As shown in figure 3.21 (c), the actual field pattern around the induction heater is complicated and difficult to simulate with the coil model of EN50366 that defines the field intensity direction only for one single axis (Y) while the maxima of the magnetic field strength shows good agreement.

Figure 3.22 shows the magnetic field strength $|H|$ [dB A/m] along the Y axis, which is defined to be in front of the source models (see figure 3.21) and is located in the same direction as the magnetic field intensity (main beam) of the coil models (r_{coil} = 0.01 and 0.07 m). As shown in this result, both coil models show good agreement with the result of the equivalent source model with the maximum difference of 3 dB.

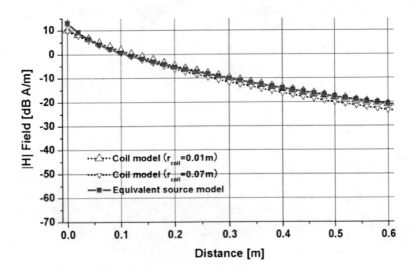

Figure 3.22: Comparison of magnetic field strength of the equivalent source model and coil model along the Y axis (induction heater).

Figures 3.23 and 3.24 show the two main field strength vectors along the Y axis (same as figure 3.22) for the coil model and the equivalent source model, respectively. As shown in figure 3.23, the dominant field strength vector is Hy as expected for the coil model. There exists for both sizes of coil models (difference less than 1 dB) a good agreement between this dominant vector Hy and the field strength ($|H|$) as shown in the figure 3.22. On the other hand, the dominant field strength vector direction (Hz) of the equivalent source model (see figure 3.24) differs from that of the coil model due to the arrangement of the eddy heating coil (spiral coil) in the induction heater, which is orientated in the Z-direction. Furthermore, the variation in the recessive field strength vector between the coil model (Hx) and the equivalent source model (Hy) differs considerably. As shown in figure 3.24, the field strength vector Hy of the equivalent source model is approximately at the same level as that of the field strength vector Hz, particularly at a small distance (0.0 m < Distance < 0.05 m). These results explain that two main field strength vectors (Hy and Hz) occur near the region of the induction heater, which significantly differ from those generated by the coil model. On the other hand, when the distance (Distance > 0.05 m) is increased, the field strength vector Hy decreases rapidly and the field strength vector Hz remains a dominant field strength vector. Thus, the field strength (|H|) shows a good agreement between both source models; however, the directions of the dominant field strength vectors differ significantly.

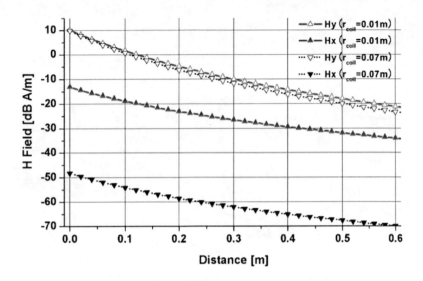

Figure 3.23: Magnetic field strength vectors of the coil models along the Y axis (induction heater).

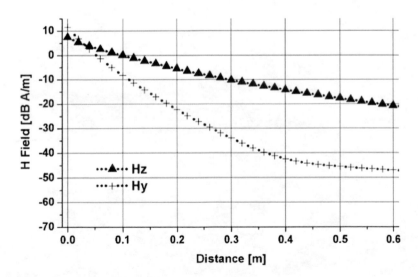

Figure 3.24: Magnetic field strength vectors of the equivalent source model along the Y axis (induction heater).

Chapter 4

Dosimetric studies on human body models

4.1 Validation of the induced current density calculation

4.1.1 Analytical calculation

The scalar potential finite difference (SPFD) method described in section 2.2.2 is applied for the numerical calculation of the induced current density in the human body model. In order to validate the numerical results, the induced current density of a six layered sphere model is compared with that obtained by the analytical results given in [41]. Figure 4.1 shows the six layered sphere model used to simulate the human head constructed with a cell size of 0.04 cm (cubic). The conductivity and the radius of each layer is described in table 4.1. The magnetic dipole moment (mz = 0.291 Am^2) is located in the front of this sphere model at a distance of 1.5 cm. Figure 4.2 shows the numerical and analytical results taken from [41] (interval of 0.02 cm) along the positive Y axis (0 to 10 cm: center to surface in the sphere model) which is calculated at 50 Hz. A good agreement between the numerical (SPFD) and the analytical results can be observed and the maximum difference between both results is 3% appearing at the peak value. The good agreement validates the numerical result.

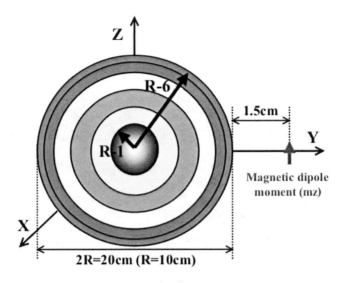

Figure 4.1: Six layer model.

Layer Number	Radius [cm]	Conductivity [S/m]
1	$0 < R \leq 8.875$	0.08
2	$8.875 < R \leq 9.125$	2
3	$9.125 < R \leq 9.250$	0.5
4	$9.250 < R \leq 9.750$	0.045
5	$9.750 < R \leq 9.857$	0.04
6	$9.857 < R \leq 10.0$	0.1

Table 4.1: Parameters of six layered sphere model.

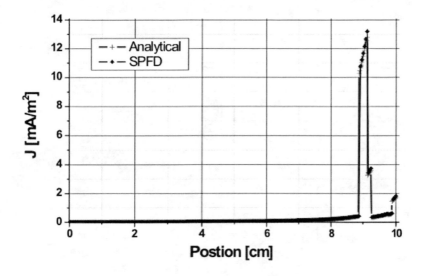

Figure 4.2: Numerical and analytical results.

4.1.2 Numerical calculation

Human body model

Figure 4.3 shows an axially symmetric homogeneous human body model ([35],[36],[37]), which is used as a standard human body model in EN50366 [5] and also applied as a numerical validation model in this section. The geometry-based numerical model of the current loop (for reference see figure 3.1) and the equivalent source model are located in front of the human body model (negative Y axis direction) at a distance d [m] from the body surface. For both source models, the maximum magnetic flux density point (Theta = 90 deg) on the cylinder surface, shown in figure 3.4 (section 3.1.3), is arranged to lie along the Y axis and to have the same height as the breast position (located on the surface of the body and 1.0 m from the bottom of the human body). A constant value of 0.1 [S/m] at 50 Hz is taken as the electrical conductivity, and the body consists of cubic voxels (size: 0.3 cm). The calculated induced current density is averaged over a cross-section area of 1 cm^2 [1].

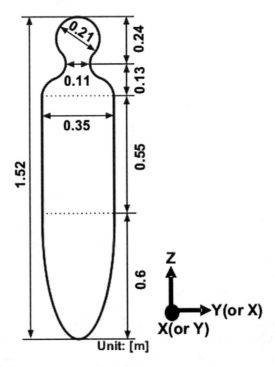

Figure 4.3: Axially symmetric homogeneous human body model.

Results

Figure 4.4 shows the distributions of the induced current density strength (J) in a spatial view from the front side and the XY-plane (Z at maximum J) for distances d (0.0, 0.05 and 0.3) m. The J distribution is calculated for both the current loop and equivalent source model, and it is found that the distributions agree very well as mentioned in the previous section with regard to the magnetic field patterns. Therefore, only the distribution of the equivalent source model is presented hereafter. As given from the J distribution in the XY-plane, the maximum induced current density (Jmax) appears at the surface of the body model and the Jmax value decreases with increasing distance. From the spatial view figures, J is found to be concentrated within a semicircular area for the small distances of d = 0 and 0.05 m. This peculiar J distribution for small distances is caused by the main beam effect in which the magnetic field intensity is concentrated along the oblique direction as described in section 3.1.3 (see figure 3.4). These results confirm that the equivalent source model maps the magnetic field characteristics to the induced current density with high accuracy. Figure 4.5 shows the influence of distance d [m] on the maximum induced current density Jmax, which occurs at the surface of the human body model as described above. In this case, the results of the current loop and the equivalent source model are compared. Good agreement is achieved again between both results, the maximum difference being only 0.8%. Further, the validity of the equivalent source model is verified for the dosimetric calculations.

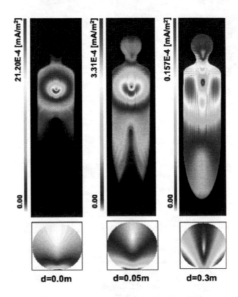

Figure 4.4: Induced current density distribution (spatial view and XY-Plane: Z at J_{max}).

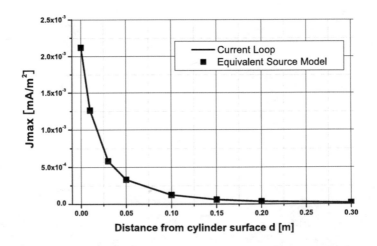

Figure 4.5: Maximum induced current density against distance.

4.2   Numerical human body models

Figure 4.6 shows two homogeneous numerical body models, which are applied for the induced current density calculations. An ellipsoidal (minor diameter = 0.4 m, major diameter = 1.8 m), a cuboid (height = 1.8 m, side length = 0.4 m) and an anatomical shape (realistic shape model with constant conductivity) are used as homogeneous body models. The conductivity for these homogeneous body models is set to 0.1 S/m. Figure 4.7 shows the anatomical realistic human body model. This model is based on cross-sectional anatomical diagrams of the visible human project [38] and is representative of a human male. The body model is constructed using 39 different tissues [39] and the electrical properties of these tissue types are often frequency dependent. In this study, the electrical properties of these tissue parameters are applied from the published information [40], which is indicated in appendix (A-1). Both the homogeneous human body model and the anatomical realistic human body model are constructed with cubic cells, with a side length of 0.3 cm.

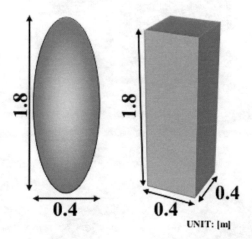

Figure 4.6: Size of homogeneous ellipsoidal and cuboid human body model.

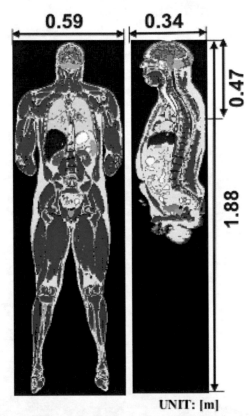

UNIT: [m]

(a) Cross-sections of whole body

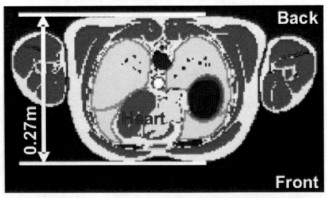

(b) Cross-section at breast (height of the heart)

Figure 4.7: Size of the anatomical human body model.

4.3   Induced current density within the body models

4.3.1   Arrangement of the body models and the test appliance

Figure 4.8 shows the arrangement of the body models and of the equivalent source model, the latter being used for the dosimetric investigations in this section. A hand mixer (see figure 3.10) is used as a test appliance for which the measured magnetic flux density level exceeds the reference level by approximately three times (see figure 3.11 (a)), as described in section 3.3.1. The equivalent source model is constructed from 1007 elementary magnetic dipoles and has a diameter of 0.22 m and a height of 0.5 m, as shown in table 3.1.

The equivalent source model is arranged in the front of the body models in order to direct the maximum magnetic flux density direction ($|B|_{max}$: Theta = 90 deg) towards the body (see figure 4.9) that is postulated to coincide with the Y axis shown in figure 4.8. Furthermore, the peak ($|B|_{max} = 355 \ \mu T$) of the equivalent source model (H = 0 cm: see figure 3.11 (b)) is located at the center height of the ellipsoidal and cuboid body models. As shown in figure 4.9, the body trunk is considered to be the exposed body part, which includes significant organs and tissues such as the heart and the central nervous system, [48] for the anatomical human body model (homogeneous and realistic). Two types of exposure cases are considered in this study. One case of the actual use of the appliance where the magnetic field being emitted towards the front of the trunk and the other case is that where the appliance is placed on the back of the body and the magnetic field is emitted towards the back of the trunk. In order to simulate these conditions, the equivalent source model is arranged both in front and at the back of the trunk. Furthermore, the breast (height of the heart) and the stomach height positions are selected for the positionning of the source model. In this section 4.3 the induced current density calculations are carried out at the 50 Hz frequency.

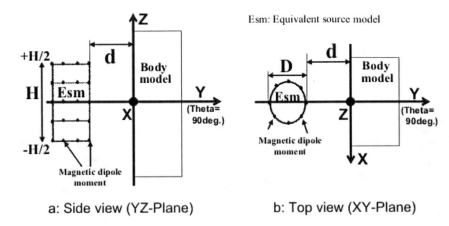

a: Side view (YZ-Plane)   b: Top view (XY-Plane)

Figure 4.8: Arrangement of the body model and the equivalent source model.

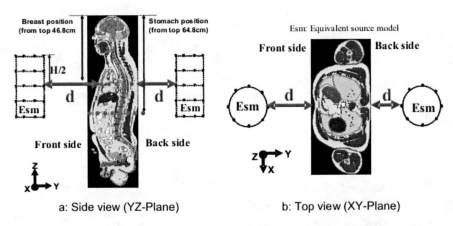

a: Side view (YZ-Plane)   b: Top view (XY-Plane)

Figure 4.9: Position of the equivalent source model arrangement for the anatomical human body model.

4.3.2 Maximum of the induced current density

Figures 4.10 and 4.11 show the results for the anatomical realistic human body model and the homogeneous body models, respectively. In these figures, the X and the Y axis represent the distance d [cm] between the source and the surface of the body model and the maximum induced current density strength Jmax [mA/m^2] averaged over 1.0 cm^2 [1], respectively. Due to the uneven surface of the front side of the anatomical human body model, the shortest distance selected is 4 cm. As shown by the results, Jmax decreases with distance for both body models as expected. If the averaged values are compared, it is found that the values for the anatomical realistic body model are approximately twice as high as those for the homogeneous body model. This difference is caused by the different conductivity parameters and complex arrangement of the tissues inside the anatomical realistic human body model. Further, the maximum current density of the anatomical realistic human model in figure 4.10 is compared with the basic restriction (ICNIRP limit for general public: 2 mA/m^2 at 50 Hz). It is shown that the maximum value of Jmax appears at the front (breast, d = 4 cm) and the value (0.37 mA/m^2) is evidently lower than the basic restriction. Hence, it is confirmed that the appliance satisfies the basic restriction although the magnetic flux density exceeds the reference level (ICNIRP limit for general public: 100 μT at 50 Hz) by approximately a factor of 3.

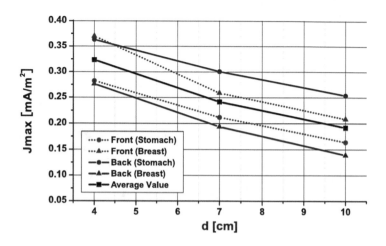

Figure 4.10: Maximum induced current density of human body model (anatomical realistic human body model).

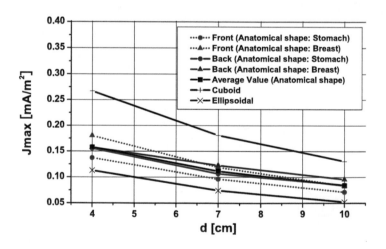

Figure 4.11: Maximum induced current density of human body model (homogeneous human body models).

4.3.3 Distribution of the induced current density

Figures 4.12, 4.13 and 4.14 show the cross-section (XY-Plane) of the averaged induced current density strength over 1.0 cm^2 in the body model at the distance of 4 cm for homogeneous human body models and the anatomical realistic human body model, respectively. The cross section is selected at the body height where the maximum induced current density strength (Jmax) occurs. As shown in the induced current density strength distribution in figures 4.12 and 4.13, the maximum induced current density strength appears at the surface of the homogeneous body models. This position is at the shortest distance and in the height range of Z = ±5 cm with respect to the source model. However, the maximum induced current density strength in figure 4.13 (d) appears at the small gap (underarm) although this does not correspond to the shortest distance (more than 10 cm) from the source model. On the other hand, Jmax occurs inside the anatomical realistic body model (3 to 7 cm from the surface) and not at the close surface of the body model (skin or fat tissue), as shown in figure 4.14. Furthermore, Jmax in the body models always appears on the right side of the source model. This is caused by the main beam (direction of the magnetic field intensity) of the test appliance that points to an oblique direction (see figure 3.17 (b) in section 3.3.4) and not the vertical (Y axis) direction. Hence, it is shown that Jmax always appears at the surface of the homogeneous body model, while it appears inside the anatomical realistic body model.

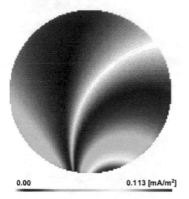 

| 0.00 | 0.113 [mA/m^2] | 0.00 | 0.267[mA/m^2] |

Figure 4.12: Induced current density distribution in the ellipsoidal and cuboid homogeneous human body model (XY-Plane: Z at Jmax).

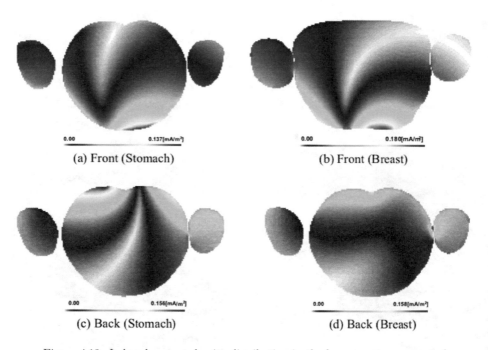

(a) Front (Stomach)

(b) Front (Breast)

(c) Back (Stomach)

(d) Back (Breast)

Figure 4.13: Induced current density distribution in the homogeneous anatomical shape human body (XY-Plane: Z at Jmax).

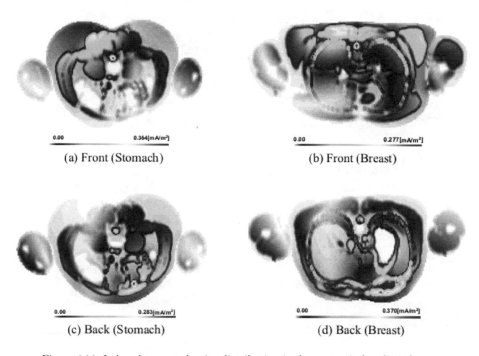

Figure 4.14: Induced current density distribution in the anatomical realistic human body (XY-Plane: Z at Jmax).

4.4   Dosimetric study using the coil source model

4.4.1   Alignment of the sources in the vicinity of human body models

Figures 4.15 (a) and (b) show the positions of the sources (for the coil and the equivalent source model) that are located in front of the axially symmetric homogeneous body model (see figure 4.3) and the anatomical realistic human body model (see figure 4.7) with distance d [m], respectively. Both body models are constructed out from cubic voxels (size: 0.3 cm), and the induced current density is calculated by the SPFD method.

The center height of the equivalent source model (H/2) and the coil model are located at the breast position, 0.52 m and 0.47 m from the top of the homogeneous and anatomical realistic body model, respectively, i.e., at the center height of the heart, respectively. For the electrical conductivity of the homogeneous body model, a constant value of 0.1 S/m [5] is applied for frequencies of 50 Hz and 21 kHz. Furthermore, 39 different tissue parameters of the anatomical realistic human body model, which depend on the frequency variation, are given in the appendix (A-1).

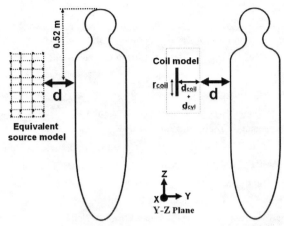

(a) Sources applied to the homogeneous body model

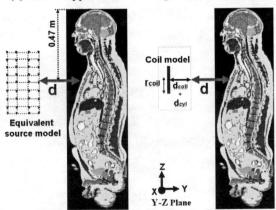

(b) Sources applied to the anatomical realistic body model

Figure 4.15: Positions of the sources applied to the human body models.

4.4.2   Results of induced current density in the body models

At low frequency of 50 Hz

In the first part of this section, a hand mixer and a drilling machine are used as the test appliances (see figure 3.10) and induced current density is investigated at 50 Hz. The figures 4.16 and 4.17 show the calculated induced current density strength values obtained from the axially symmetric homogeneous human body model and the anatomical realistic human body model, respectively. In these figures, the X and the Y axis represent the distance [m] between the source (coil model: ■, equivalent source model: ×) and the surface of the body model. The maximum induced current density strength J_{max} [mA/m^2] is averaged over 1.0 cm^2 [1]. Due to the uneven surface on the front of the anatomical body model, the minimum distance is selected to be 0.04 m. As shown in figure 4.17, J_{max} of the anatomical body model is approximately three to four times higher than that of the homogeneous body model (figure 4.16) in the small distance range, primarily due to the different conductivity parameters. When the results between the coil model and equivalent source model are compared, J_{max} of the coil model was found to be higher for both appliances and body models. A difference factor of approximately 1.6 is observed in case of drilling machine for both the homogeneous and the anatomical body model, while in case of the hand mixer, it is approximately 1.1 for both the homogeneous and anatomical body model. These differences could result from the disparity in the dominant magnetic field vector between both source models as described in section 3.3.4. However, this disparity in the field vector has caused no significant difference in the induced current density values. Hence, despite the simple construction of the coil model that is prescribed in EN50366, a fair coincidence is observed between both source models (factor of 1.6). Furthermore, it is observed that the induced current density value obtained for the coil model is larger when compared to the real appliance (i.e. the equivalent source model).

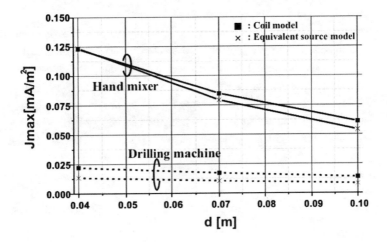

Figure 4.16: Maximum induced current density against distance (axially symmetric homogenous human body model).

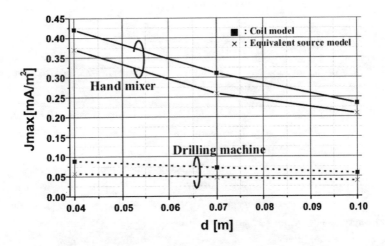

Figure 4.17: Maximum induced current density against distance (anatomical realistic human body model).

At intermediate frequency of 21 kHz

In the second part of this section, the induction heater is used as a test appliance (see figure 3.13) and the induced current density is investigated at 21 kHz. As already described in section 3.3.1, the measured peak spectrum of the magnetic field is in the range between 20 kHz and 22 kHz. Hence, the variation in the induced current density against the frequency variation is investigated in the first step. Table 4.2 shows the maximum induced current density for the coil model (r_{coil} = 0.01 m) and the equivalent source model, respectively. The distance between the body model and source models is 0.04 m and the frequency varies at 20 kHz, 21 kHz and 22 kHz, which ranges between ±5% with respect to the center frequency of 21 kHz. As shown in the both source model, the induced current density value varies in the same manner as the frequency, i.e. ±5%. Similar results are discussed for the homogeneous field exposure in reference [53], which corroborates the validity of these calculated results. Consequently, the calculated induced current density varies linearly with the frequency of the exposed magnetic field. Hence, in this section, the induced current density calculated using the equivalent source model involves an error range, which can be derived from the frequency variation range of the measured magnetic field.

	Coil model (r_{coil} = 0.01m)	Equivalent source model
20 kHz	2.23 [mA/m^2]	0.572 [mA/m^2]
21 kHz	2.34 [mA/m^2]	0.601 [mA/m^2]
22 kHz	2.45 [mA/m^2]	0.629 [mA/m^2]

Table 4.2: Maximum induced current density against the variation of frequency (d = 0.04m).

Figure 4.18 shows the calculated maximum induced current density strength Jmax in the axially symmetric homogeneous human body model against the distance at the center frequency of 21 kHz, for the coil models (r_{coil} = 0.01 and 0.07 m) and the equivalent source model, respectively. As described above, these Jmax values involve an error of approximately ±4% that arises from the frequency variation (see section 3.3.1) in the measured magnetic field results. Comparing the results between both source models, Jmax is larger for either coil model. The difference between the coil model and the equivalent source model is approximately given by a factor of 2.2 and 2.7 for the coil sizes of r_{coil} = 0.07 m and r_{coil} = 0.01 m, respectively. Furthermore, the same procedure is conducted for the anatomical realistic human body model, and the results are shown in figure 4.19. Similar to the homogeneous body model, Jmax is larger for either coil model. The differences between the coil model and the equivalent source model is approximately given by a factor of 2.1 and 2.4 for the coil sizes of $r_{coil} = 0.07$ m and $r_{coil} = 0.01$ m, respectively.

Hence, the coil model produces higher current density compared to the induction heater. It is also shown that the smaller size of the coil model leads to larger Jmax values, even though the value G remains constant. The large difference between both source models could be caused by the different dominant field strength vector of the coil model (Hy) and the induction heater (Hz), as described in section 3.3.5. Consequently, the application of the coil model prescribed in EN50366 leads to the worst case assumption of the induced current density for the applied induction heater.

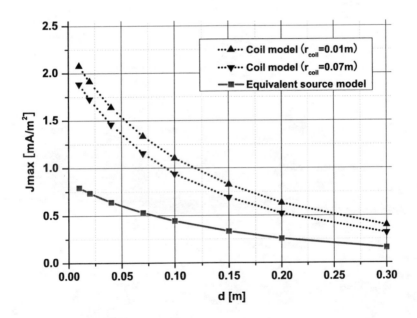

Figure 4.18: Maximum induced current density Jmax against distance (axially symmetric homogeneous human body model).

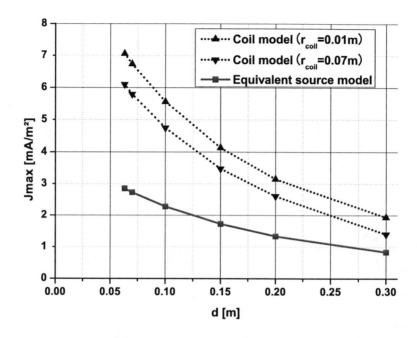

Figure 4.19: Maximum induced current density Jmax against distance (anatomical realistic human body model).

Chapter 5

Conclusions

An equivalent source model that allows efficient low frequency dosimetric investigations has been proposed. This model enables the reproduction of complicated inhomogeneous magnetic field characteristics associated with real electrical appliances with a complete three-dimensional representation of the magnetic vector field (field components and directions). In the first step, the validity of the equivalent source model is numerically investigated by evaluating the accuracy of the magnetic field distribution around the numerical reference model of a current loop. Almost identical magnetic field strengths, directions of the field vectors (polarization), and directions of the field intensities (main beam) are obtained for both models, with a maximum variation of less than 1%. Furthermore, the validity of the induced current density, which is used in the European norm EN50366, is investigated for a homogeneous human body model. The results obtained from the equivalent source model and the current loop are almost identical. The excellent agreement confirms the validity of the equivalent source model for both magnetic field distribution and induced current density in the human body model. In the second step, the validity of the equivalent source model is verified for commercial appliances, comparing the results of the measured and calculated magnetic flux density around the appliances. A good agreement is observed between the measured and calculated results with a maximum difference of approximately 10%, and the validity of the equivalent source model is also confirmed for real household appliances (i.e. practical application).

By applying this equivalent source model, the inhomogeneous field dosimetry at a low frequency of 50 Hz is calculated with the SPFD method, and the dosimetric differences between the numerical human body models are investigated using a

commercial hand mixer, in which the level of the leakage magnetic field exceeds
the reference level. In the case of the numerical human body models, the anatomical realistic one as well as the simple human body models (with homogeneous
conductivity), which have already been used in previous studies, are used for this
investigation. The resulting current densities are compared with the basic restrictions (ICNIRP guideline) and confirmed by applying the anatomical body model
such that the maximum values of the induced current density satisfy these limits
for a typical exposition position, even if the magnetic flux density value exceeds approximately three times that of the reference level. Furthermore, it is observed that
the maximum induced current density appears at the surface with the homogeneous
body models, while it appears inside the body model for the anatomically realistic
case.

Finally, the magnetic field properties and the dosimetry at a low frequency of 50
Hz caused by two commercial appliances (drilling machine and hand mixer) are
investigated with the coil model and with the equivalent source model as a reference.
In the case of either test appliances, the coil size is derived from the prescribed
procedures of European standard EN50366. The accuracy of the magnetic field
vectors and the induced current density is investigated by comparison with the
results of the real appliance (i.e. the equivalent source model). A good agreement
for the dominant magnetic field vector is observed on the axis (maximum difference
of 5 dB) that has the same direction as the field intensity as calculated with the
coil model. These results confirm the reasonable accuracy for the dominant field
vector (direction and magnitude) obtained by the coil model. On the other hand,
the calculated induced current density values in the human body models show an
agreement with the maximum difference, given by a factor of 1.6 (anatomical body
model) as compared with the real appliance. Furthermore, it is shown that the
values of both the magnetic field and the induced current density obtained from
the coil model are higher than those of the test appliances. The magnetic field
properties and the dosimetry at an intermediate frequency of 21 kHz are investigated
using a commercial induction heater in a similar manner as described above. A
good agreement between the coil model and the induction heater is confirmed with
respect to the magnetic field strength. On the other hand, the dominant field
vector differs significantly due to the different arrangements of the eddy heating
coil (located under the vessel in the induction heater body) and the coil model

prescribed in EN50366, which causes the different field intensity directions. The maximum induced currents in the body models are also investigated for both the coil model and the induction heater. The values obtained for the coil model are three times larger than those for the real induction heater (up to a factor of 2.4 and 2.7 for homogeneous and anatomical body model, respectively). This difference could be caused by the differences in the dominant field vector, which are described above. With these results, a good agreement between the coil model and the test appliances is confirmed for the magnetic field strength on the investigated single axis. On the other hand, it is observed that, even with this good agreement for the magnetic field strength, a significant difference in the dominant field vector led to a large difference in the induced current density of the body model.

Bibliography

[1] ICNIRP, Guidelines for limiting exposure to time-varying electric magnetic and electromagnetic fields (up to 300 GHz), Health Physics, Vol. 41, No. 4, pp. 449-522, 1998.

[2] J.R. Gauger, Household appliance magnetic field survey, IEEE Transaction on Power Apparatus and Systems, Vol. PAS-104, No.9, pp. 2436-2444, 1985.

[3] European Commission, Standardisation Mandate Addressed to CEN, CENELEC and ETSI in the Field of Electrotechnology, Information Technology and Telecommunications, M/305 EN, 2000.

[4] S. Niinisto, Council Recommendation of 12 July 1999 on the limitation of exposure of the general public to electromagnetic fields (0 Hz to 300 GHz), Official Journal of the European Communities, L 199/59, July 1999.

[5] European standard EN50366, Household and similar electrical appliances - Electromagnetic fields - Methods for evaluation and measurement, CENELEC, 2003.

[6] P. Baraton, and B. Hutzler, Magnetically induced currents in the human Body, IEC Technology Assessment, 1995.

[7] S. Tofani, P. Ossola, G.d' Amore, and O.P. Gandhi, Electric Field and Current Density Distributions Induced in an Anatomically-Based Model of the Human Head by Magnetic Fields from a hair Dryer, Health Physics, vol. 68, no. 1, 1995, pp. 71-79.

[8] J. Cheng, M. A. Stuchly, C.D. Wagter, and L. Martens, Magnetic field induced currents in a human head from use of portable appliances", Physics in Medicine and Biology, vol. 40, pp.495-510, 1995.

[9] D. L. Mader and S. B. Peralta, Residental Exposure to 60 Hz Magnetic fields from Appliances, Bioelectromagnetics 13:287-301, 1992.

[10] L. Zaffanella, T. Sullivian, and I. Visintainer, Magnetic field characterization of electrical appliances as point sources through in situ measurement, IEEE Transaction on Power Delivery, Vol. 12, No. 1, pp.402-406, 1997.

[11] K. Yamazaki, and T. Kawamoto, Simple estimation of equivalent magnetic dipole moment to characterize ELF magnetic fields generated by electric appliances incorporating harmonics, IEEE Transactions on EMC, Volume 43, Issue 2, pp.24 -245, 2001.

[12] K. Wake, T. Tanaka, and M. Taki, Analysis of induced currents in a rat exposed to 50 Hz linearly and circularly polarized magnetic fields, Bioelectromagnetics, Volume 21, Issue 5, pp.354-363, 2000.

[13] N. Kuster, and Q. Balzano, Experimental and Numerical Dosimetry, In Mobile Communications Safety (eds. N. Kuster, Q. Balzano, and J.C. Lin), pp. 13-58, Chapman and Hall, London, UK, 1997.

[14] R. W. Y. Habash, Electromagnetic fields and radiation (Human Bioeffects and Safety), Marcel Dekker Inc., 2002.

[15] H. D. Durney, and D. A. Christensen, Basic Introduction to Bioelectromagnetic, CRC Press, Boca Raton, FL, 1999.

[16] National Research Council, Possible Health Effects of Exposure to Residental Electric and Magnetic Fields, National Academy Press, 1997.

[17] K. R. Foster, and H. P. Schwan, Dielectric permittivity and electrial conductivity of biological materials, Handbook of Biological Effects of Electromagnetics Fields, C. Polk and E.Postow, eds. Boca Raton, Fla.: CRC Press, pp. 27-96, 1986.

[18] C. M. Furse, and O. P. Gandhi, Calculation of Electric Fields and Currents Induced inn a Millimeter-Resolution Human Model at 60 Hz Using the FDTD Method, Bioelectromagentics 19, pp.293-299, 1998.

[19] W. Xi, M. A. Stuchly, and O. P. Gandhi, Induced electric currents in Models of Man and Rodents from 60 Hz Magnetic Fields, IEEE Transaction on Biomedical Engineering, Vol. 41, No.11, 1994.

[20] R. Meyer, In Vitro Experiments dealing with the Biological Effects of RF Fields at Low Energies, COST 244 Projects, Forum on Future European Research on Mobile Communications and Health, pp. 19-20, 1999.

[21] J. E. Moulder, Biological Studies of Power-Frequency Fields and Carcinogenesis, IEEE Engineering in Medicine and Biology 15, pp. 31-40, 1996.

[22] T. S. Tenforde, Biological Interactions and Potential Health Effects of Extremely Low Frequency Magnetic Fields from Power Lines and Other Common Sources, Annual Review of Public Health 13, pp. 173-196, 1992.

[23] L. E. Anderson, and W. T. Kaune, Electric and Magnetic Fields at Extremely Low Frequencies, In Nonionizing Radiation Protection (eds. Suess, M. J., and D. A. Benwell-Morison), World Health Organization Regional Publications, European Series 25, pp. 175-243, 1989.

[24] J. P. Reilly, Peripheral Nerve Simulation by Induced Electric Currents, Exposure to Time-Varying Magnetic Fields, Medical and Biological Engineering and Computing 3, pp. 101-109, 1989.

[25] C. T. Tai, Generalized vector and dyadic analysis, Applied Mathematics in Field Theory, IEEE Press, 1991.

[26] F. M. Landstorfer, Hochfrequenztechnik III, Lecture material, Chapter 1.5.

[27] R. F. Harrington, Time-Harmonic Electromagnetic Fields, The IEEE Press Series on Electromagnetic Wave Theory, 2001.

[28] R. F. Harrington, Introduction to Electromagnetic Engineering, Dover Publications, INC., 2003.

[29] G. Lehner, Elektomagnetische Feldtheorie für Ingenieure und Physiker, Springer, 1996.

[30] J. D. Jackson, Classical Electrodynamics-Third Edition, John Wiley & Sons Inc., 1998.

[31] S. Nishizawa, W. Spreitzer, H.-O. Ruoß, and F. M. Landstorfer, Equivalent Source Model for Electrical Appliances emitting low Frequency Magnetic Fields, Proceeding of 31th European Microwave Conference 2001, Vol.3, pp.117-120, September 2001.

[32] Software tool FEKO: http://www.feko.info.

[33] U. Jakobus, Erweiterte Momentenmethode zur Behandlung kompliziert aufge-bauter und elektrisch großer elektromagnetischer Streuprobleme, Fortschritts-berichte VDI, Reihe 21, Nr.171, VDI Verlag, Düsseldorf, 1995

[34] U. Kampet, and W. Hiller, Measurement of magnetic flux densities in the space around household appliances, Proceedings of NIR 99, Nichtionisierende Strahlung, 31. Jahrestagung des Fachverbandes für Strahlenschutz, Köln, vol. II, pp. 885-891, 1999.

[35] Deutsch Industrie Norm (DIN) / German Industrial Norm, 33402 Teil 2.

[36] L. Geisbusch, U. Jakobus, F. M. Landstorfer, M. Maier, H.-O. Ruos, W. Spre-itzer, and J. Waldmann, Worst case study of the coupling between electrodes of cardiac pacemakers and transmitting antennas, Proceeding of 22nd Annual Bioelectromagnetics Society Meeting, München, pp. 230-231, 2000.

[37] F. M. Landstorfer, Development of a model describing the coupling between electrodes of cardiac pacemakers and transmitting antennas in their close vicinity in the frequency range from 50 MHz to 500 MHz, Final report (ID number 808), Forschungsgemeinschaft Funk (FGF), 1999.

[38] National Library of Medicine: The Visible Human Project http://www.nlm.nih.gov/research/visible/visible_human.html

[39] Air Force Research Laboratory, AFRL/HEDR, ftp://starview.brooks.af.mil/EMF/dosimetry_models/

[40] Italian National Research Council, Calculation of the dielectric prop-erties of Body Tissues in the frequency range 10 Hz - 100 GHz, http://sparc10.iroe.fi.cnr.it/tissprop/htmlclie/htmlclie.htm

[41] Y. Kamimura, M. Kojima, and Y. Yamada, Induced Current Inside a Sphereri-cal Model of a Human Head Due to Magnetic Current Sources of AC Drive Electric Shaver, IEICE, B, Vol.J85-B, No.5, pp.706-714, 2002.

[42] T. W. Dawson, J. D. Moerloose, and M. A. Stuchly, Comparison of magnetically induced ELF fields in humans computed by FDTD and scalar potential FD codes, Applied Computational Electromagnetics Society (ACES) Journal, Vol. 11, No. 3, pp 63-71, 1996.

[43] T. W. Dawson, and M. A. Stuchly, Analytic validation of a three-dimensional scalar-potential finite-difference code for low-frequency magnetic induction, Ap-plied Computational Electromagnetics Society (ACES) Journal, Vol. 11, No. 3, pp 72-81, 1996.

[44] P. J. Dimbylow, Current densities in a 2 mm resolution anatomically realistic model of the body induced by low frequency electric fields, Physics in Medicine and Biology, 45, No-4, pp.1013-1022, 2000.

[45] J. Van. Blandel, Electromagnetic Fields, Washington D.C, revised printing edi-tion, Hemisphere Publishing Corporation, Washinton DC, 1985.

[46] T. W. Dawson and M. A. Stuchly, High-Resolution Organ Dosimetry for Human Exposure to Low-Frequency Magnetic Fields, IEEE Transaction on Magnetics, Vol. 34, No.3, pp.708-718, May 1998.

[47] T. W. Dawson, K. Caputa, and M. A. Stuchly, Influence of human model resulution on computed currents induced in organs by 60 Hz magnetic fields, Bioelectromagentics 18: pp.478-490, 1997.

[48] R. Matthes, Response to questions and comments on ICNIRP - Guidelines for limiting exposure to time-varying electric, magnetic and electromagnetic fields (up to 300 GHz), Health Physics, vol.75, No.4, pp. 438-439, 1998.

[49] S. Nishizawa, H.-O. Ruoss, F. Landstorfer, and O. Hashimoto, Numerical study on an equivalent source model for inhomogeneous magnetic field dosimetry in the low frequency range, IEEE Transactions on Biomedical Engineering, Vol.51, No.4, pp. 612-618, 2004.

[50] W. H. Press, S. A. Teukolsky, W. T. Vetterling, and B. P. Flannery, Numerical Recepies in C (The art of Scientific Computing - Second Edition, Cambridge University Press, 1992.

[51] A. Meister, Numerische linearer Gleichungsststeme, Vieweg, 1999.

[52] H. R. Schwarz, Numerische Mathematik,B.G. Teubner Stuttgart, 1997.

[53] A. Guy, S. Davidow, G. Yang and C. Chou, Determination of Electric Current Distributions in Animals and Humans Exposed to a Uniform 60 Hz high intensity electric Field, Bioelectromagnetics, vol. 3, pp.47-71, 1982.

[54] J. E. Moulder, Power Lines and Cancer FAQs, Electromagnetic Fields and Human Health, Medical College of Wisconsin, 1999.

[55] T. A. Litovitz, D. Krause, and J. M. Mullins, Effect of Coherence Time of the Applied Magnetic Field on Ornithine Decarboxylase Activity, Biochemical and Biophysical Research Communication 178, pp. 862-865, 1991.

[56] J. Jutilainen, and A. Limatainen, Mutation Frequency in Salmonella Exposed to Weak 100 Hz Magnetic Fields, Hereditas 104, pp. 145-147, 1986.

[57] A. Cossarizza, D. Monti, G. Moschini, R. Cadossi, F. Bersani, and and C. Franceschi, DNA Repair after Gamma Irradiation in Lymphocytes Exposed to Low-Frequency Pulsed Electromagnetic Fields, Radiation Research 118, pp. 161-168, 1989.

[58] M. E. Fraizier, J. A. Reese, J. E. Morris, R. F. Jostes, and D. L. Miller, Exposure of Mammalian Cells to 60 Hz Magnetic and Electric Fields: Analysis of DNA Repair of Induced Single- Strand Breaks, Bioelectromagnetics 11, pp. 229-239, 1990.

[59] M. R. Scarfi, F. Bersani, A. Cossarizza, D. Monti, G. Castellani, R. Cadossi, G. Franceschetti, and C. Franceschi, Spontaneous and Mitomycin-C-Induced Micronuclei in Human Lymphocytes Exposed to Extremely Low Frequency Pulsed Magnetic Fields, Biochemical and Biophysical Research Communication 176, pp. 194-200,1991.

[60] G. K. Livingston, K. L. Witt, O. P. Gandhi, I. Chaterjee, and J. Roti, Reproductive Integrity of Mamalian Cells Exposed to Power Frequency Electromagnetic Fields, Environmental and Molecular Mutagenesis 17, pp. 49-58, 1991.

[61] D. D. Ager, and J. A. Radul, Effect of 60 Hz Magnetic Fields on Ultraviolet Light-Induced Mutation and Mitotic Recombination in Saccharomyces Cerevisaie, Mutation Research 283, pp. 279-286, 1992.

[62] D. E. Hintenlag, Synergistic Effects of Ionizing Radiation and 60 Hz Magnetic Fields, Bioelectromagnetics 14, pp. 545-551, 1993.

[63] C. D. Cain, D. L. Thomas, and W. R. Adey, 60 Hz Magnetic Field Strength Dependency and TPA-Induced Focus Formation in Co-Cultures of C3H/10T/2 Cells, Annual Review of Research on Biological Effects of Electric and Magnetic Fields, pp. 55, 1994.

[64] S. A. Tofani, Ferrara, L. Anglesio, and G. Gilli, Evidence for Genotoxic Effect of Resonant ELF Magnetic Fields, Bioelectrochem. Bioenerg. 36, pp. 9-13, 1995.

[65] A. Antonopoulos, B. Yang, A. Stamm, W. D. Heller, and G. Obe, Cytological Effects of 50 Hz Electromagnetic Fields on Human Lymphocytes In Vitro, Mutation Research 346, pp. 151-157, 1995.

[66] O. P. Cantoni,P. Sestili, M. Fiorani, and M. Dacha, The Effect of 50 Hz Sinusoidal Electric and/or Magnetic Fields on the Rate of Repair of DNA Single/Double Strand Breaks in Oxidatively Injured Cells, Biochemistry and Molecular Biology International 37, pp. 681-689, 1995.

[67] H. Okonogi, M. Nakagawa, and Y. Tsuji, The Effects of a 4.7 uT Static Magnetic Field on the Frequency of Micronucleated Cells Induced by Mitomycin C, Tokushima Journal of Experimental Medicine 180, pp. 209-215, 1996.

[68] M. A. Morandi, C. M. Pak, R. P. Caren, and L. D. Caren, Lack of an EMF-Induced Genotoxic Effect in the Ames Assay, Life Science 59, pp. 263-271, 1996.

[69] I. Lagroye, and J. L. Poncy, The Effect of 50 Hz Electromagnetic Fields on the Formation of Micronuclei in Rodent Cell Lines Exposed to Gamma Radiation, International Journal of Radiation Biology 12, pp. 249-254, 1997.

[70] J. Jacobson-Kram, J. Tepper, P. Kuo, R. H. San, P. T. Curry, V. O. Wagner, D. L.Putman, Evaluation of the Potential Genotoxicity of Pulsed Electric and Electromagnetic Field Used for Bone Growth Stimulation, Mutation Research 388, pp. 45-57, 1997.

[71] M. R. Scarfi, M. B. Lioi, M. DellaNoce, O. Zeni, C. Franceschi, D. Monti, and G. Castellani, Exposure to 100 Hz Pulsed Magnetic Fields Increases Micronucleus Frequency and Cell Proliferation in Human Lymphocytes, Bioelectrochemistry and Bioenergetics43,pp. 77-81, 1997.

[72] E. K. Balcer-Kubiczek, X. Zhang, L. Han, G. H. Harrison, C. C. Davis, X. Zhou, V. Loffe, W. A. McCready, J. M. Abraham, and S. J. Meltzer, BIGEL Analysis of Gene Expression in HL60 Cells Exposed to X Rays or 60 Hz Magnetic Fields, Radiation Research 150, pp. 663-672, 1998.

[73] O. N. Pakhomova, M. L. Belt, S. P. Mathur J. C. Lee, and Y. Akyel, Ultra-Wide Band Electromagnetic Radiation and Mutagenesis in Yeast, Bioelectromagnetics 19, pp. 128-130, 1998.

[74] B. I. Rapley, R. E. Rowland, W. H. Page, and J. V. Podd, Influence of Extremely Low Frequency Magnetic Fields on Chromosomes and the Mitotic Cycle in Vicia Faba L, the Broad Bean, Bioelectromagnetics 19, pp. 152-161, 1998.

[75] H. Yaguchi, M. Yoshida, Y. Ejima, and J. Miyakoshi, Effect of High-Density Extremely Low Frequency Magnetic Fields on Sister Chromatic Exchanges in Mouse m5S Cells, Mutation Research 440, pp. 189-194, 1999.

[76] A. M. Khalil, and W. Qassem, Cytogenetic Effects of Pulsing Electromagnetic Field on Human Lymphocytes In Vitro: Chromosomes Aberrations, Sister Chromatid Exchanges and Cell Kinetics, Mutation Research 241, pp. 141-146, 1991.

[77] D. W. Fairbairn, and K. L. O'Neill, The Effect of Electromagnetic Field Exposure on the Formation of DNA Single Strand Breaks in Human Cells, Cellular and Molecular Biology Letters 4, pp. 561 -567, 1994.

[78] J. Miyakoshi, Y. Mori, N. Yamagishi, K. Yagi, and H. Takebe, Suppression of High-Density Magnetic Field (400 mT at 50 Hz)-Induced Mutations by Wild-Type p53 Expression in Human Osteosarcoma Cells, Biochemical and Biophysical Research Communications 243, pp. 579-584, 1998.

[79] I. Dibirdik, D. Kristupaitis, T. Kurosaki, L. Tuel-Ahlgren, A. Chu, D. Pond, D. Tuong, R. Luben, and F. M. Uckun, Stimulation of Src Family Protein Tyrosine Kinases as a Proximal and Mandatory Step for SYK Kinase- Dependent Phospholipase C Gamma 2 activation in lymphoma B-Cells Exposed to Low Energy Electromagnetic Fields, Journal of Biological Chemistry 273, pp. 4035- 4039, 1998.

[80] D. S. Beniashvili, V. G. Biniashvili, and M. Z. Menabde, Low-Frequency Electromagnetic Radiation Enhances the Induction of Rat Mammary Tumours by Nitrosomethyl Urea, Cancer Letters 61, pp. 75-79, 1991.

[81] W. Loscher, M. Mevissen, W. Lehmacher, and A. Stamm, A Tumor Promotion in a Breast Cancer Model by Exposure to a Weak Alternating Magnetic Field, Cancer Letters 71, pp. 75-81,1993.

[82] Health Effects from Exposure to Power-Line Frequency Electric and Magnetic Fields: Prepared in Response to the 1992 Energy Policy Act (PL 102-486, Section 2118), National Institute of Environmental Health Sciences (NIEHS) and National Institutes of Health, 1999.

[83] D. D. Ager, and J. A. Radul, Effect of 60 Hz Magnetic Fields on Ultraviolet Light-Induced Mutation and Mitotic Recombination in Saccharomyces Cerevisiae, Mutation Research 283, pp. 279-286, 1992.

[84] M. A. Morandi, C. M. Pak, R. P. Caren, and L. D. Caren, Lack of an EMF-Induced Genotoxic Effect in the Ames Assay, Life Science 59, pp. 263-271, 1996.

[85] R. G. Stevens, Electric Power Use and Breast Cancer, A Hypothesis, American Journal of Epidemiology 125, pp. 556-561, 1987.

[86] D. E. Blask, S. T. Wilson, J. D. Saffer, M. A. Wilson, L. E. Anderson, and B. W. Wilson, Culture Conditions Influence the Effects of Weak Magnetic Fields on the Growth-Response of MCF-7 Human Breast Cancer Cells to Melatonin in Vitro, Annual Review of Research on Biological Effects of Electric and Magnetic Fields from the Generation, Delivery and Use of Electricity, p. 65, U.S. Dept. of Energy, Savannah, GA, 31 October to 4 November 1993.

[87] R. P. Liburdy, T. R. Sloma, R. Sokolic, and P. Yaswen, ELF Magnetic Fields, Breast Cancer and Melatonin: 60 Hz Fields Block Melatonin's Oncostatic Action

on ER+ Breast Cancer Cell Proliferation, Journal of Pineal Research 14, pp. 89-97, 1993.

[88] B. Selmaoui, and Y. Touitou, Sinusoidal 50 Hz Magnetic Fields Depress Rat Pineal NAT Activity and Serum Melatonin. Role of Duration and Intensity of Exposure, Life Sciences 57, pp.1351-1358, 1995.

[89] J. D. Harland, and R. P. Liburdy, ELF Inhibition of Melatonin and Tamoxifen Action on MCF-7 Cell Proliferation: Field Parameters, The Bioelectromagnetics Society Meeting, Victoria, British Columbia, Canada, 1996.

[90] D. Maisch, Melatonin, Tamoxifen, 50-60 Hertz Electromagnetic Fields and Breast Cancer: A Discussion Paper, Australian Senate Hansard, pp. 1-9, 1997.

[91] Computation of the dielectric properties of body tissues at RF and microwave, `http://www.brooks.af.mil/AFRL/HED/hedr/reports/dielectric/Report/Report.html`

[92] C. Gabriel, T. Y. A. Chan, and E. H. Grant, Admittance models for open ended coaxial probes and their place in dielectric spectroscopy, Physics in Medicine and Biology, 39, 12, pp.2183-2200, 1994.

[93] C. Gabriel, and E. H. Grant, Dielectric sensors for industrial microwave measurements and control, Microwellen und HF Magazin, vol.15, pp.643-645, 1989.

[94] P. Schwan, Linear and nonlinear electrode polarisation and biological materials, Annals of Biomedical Engineering: 20, pp. 269-288, 1992.

[95] C. H. Durney, H. Massoudi, and M. F. Iskander, Radiofrequency radiation dosimetry handbook, Brooks Air Force Base- USAFSAM-TR-85-73, 1986.

[96] L. A. Geddes, and L. E. Barker, The specific resistance of biological material - a compendium of data for the biomedical engineer and physiologist., Medical and Biological Engineering, 5, pp.271-293, 1967.

[97] M. A. Stuchly, and S. S. Stuchly, Dielectric properties of biological substances - tabulated, Journal of Microwave Power, 15, 1, 19-26, 1980.

[98] K. R. Foster, and H. P. Schwan, Dielectric properties of tissues and biological materials: A critical review, Critical Reviews in Biomedical Engineering, 17, 1, pp.25-104, 1989.

[99] F. A. Duck, Physical properties of tissue:A comprehensive reference book, Academic Press, Harcourt Brace Jovanovich, Publishers, 1990.

Appendix

A-1. Tissue parameters of anatomical body model

Low frequency - 50 Hz [40]

Tissue	Conductivity [S/m]	Tissue	Conductivity [S/m]
Air	0.0	Lung Deflated	0.2055
Aorta	0.2611	Lung Inflated	0.06842
Bladder	0.2054	Lymph	0.5214
Blood	0.7	Mucous Membrane	0.0004272
Blood Vessel	0.2611	Muscle	0.2333
Body Fluid	1.5	Nail	0.02005
Bone Cancelous	0.0807	Nerve	0.0274
Bone Cortical	0.02005	Oesophagus	0.5214
Bone Marrow	0.001649	Ovary	0.3214
Brain Grey Matter	0.07526	Pancreas	0.5214
Brain White Matter	0.05327	Prostate	0.4214
Breast Fat	0.02265	Retina	0.5027
Cartilage	0.1714	Skin Dry	0.0002
Cerebellum	0.09526	Skin Wet	0.0004272
Cerebroid Spinal Fluid	2.0	Small Intestine	0.5215
Cervix	0.3445	Spinal Chord	0.0274
Colon	0.05454	Spleen	0.0857
Cornea	0.4214	Stomach	0.5214
Duodenum	0.5214	Tendon	0.2698
Dura	0.5005	Testis	0.4214
Eye Sclera	0.5027	Thymus	0.5214
Fat	0.01955	Thyroid	0.5214
Gall Bladder	0.9	Tongue	0.2714
Gall Bladder Bile	1.4	Tooth	0.02005
Gland	0.5214	Trachea	0.3005
Heart	0.08273	Uterus	0.2293
Kidney	0.08924	Vacuum	0.0
Lens	0.3214	Vitreous Humor	1.5
Liver	0.03668		

Intermediate frequency - 21 kHz [40]

Tissue	Conductivity [S/m]	Tissue	Conductivity [S/m]
Air	0.0	Lung Deflated	0.2514
Aorta	0.3148	Lung Inflated	0.09741
Bladder	0.2147	Lymph	0.5315
Blood	0.7002	Mucous Membrane	0.008446
Blood-Vessel	0.3148	Muscle	0.3452
Body-Fluid	1.5	Nail	0.02052
Bone Cancelous	0.08297	Nerve	0.05436
Bone Cortical	0.02052	Oesophagus	0.5314
Bone Marrow	0.002908	Ovary	0.3327
Brain Grey Matter	0.1206	Pancreas	0.5315
Brain White Matter	0.07294	Prostate	0.4316
Breast Fat	0.02479	Retina	0.5121
Cartilage	0.1763	Skin Dry	0.0002155
Cerebellum	0.1406	Skin Wet	0.008446
Cerebroid Spinal Fluid	2.0	Small Intestine	0.5684
Cervix	0.5404	Spinal Chord	0.05436
Colon	0.2418	Spleen	0.1138
Cornea	0.4597	Stomach	0.5314
Duodenum	0.5314	Tendon	0.387
Dura	0.5015	Testis	0.4316
Eye Sclera	0.5121	Thymus	0.5315
Fat	0.02403	Thyroid	0.5315
Gall Bladder	0.9001	Tongue	0.2816
Gall Bladder Bile	1.4	Tooth	0.02052
Gland	0.5315	Trachea	0.3192
Heart	0.173	Uterus	0.5195
Kidney	0.1474	Vacuum	0.0
Lens	0.3368	Vitreous Humor	1.5
Liver	0.06126		

A-2. Frequency variation of the tissue parameters [91]

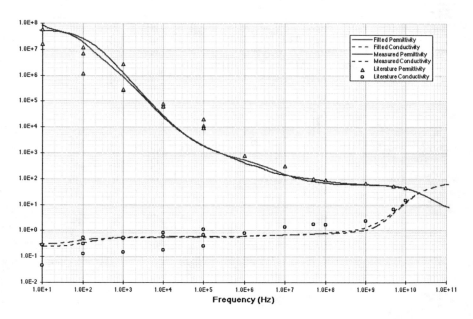

Figure 5.1: Muscle tissue (▲● Literature: [92]-[99]).

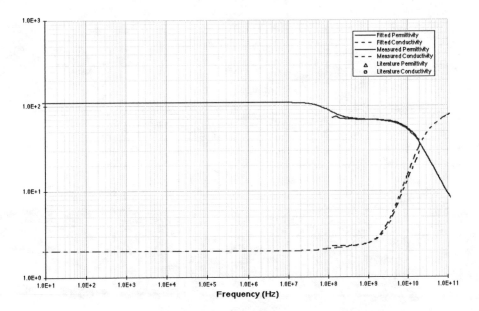

Figure 5.2: Cerebroid spinal fluid tissue (▲● Literature: [92]-[99]).

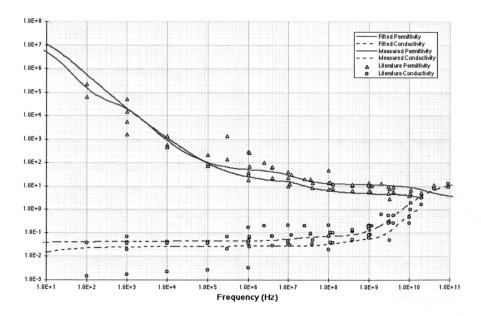

Figure 5.3: Fat tissue (▲● Literature: [92]-[99]).

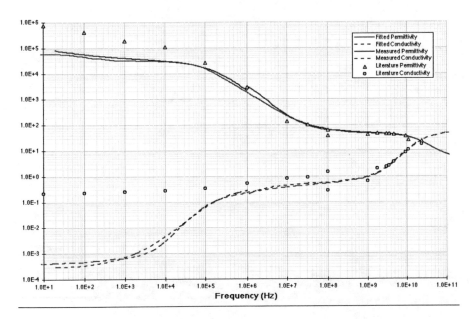

Figure 5.4: Skin (wet) tissue (▲● Literature: [92]-[99]).

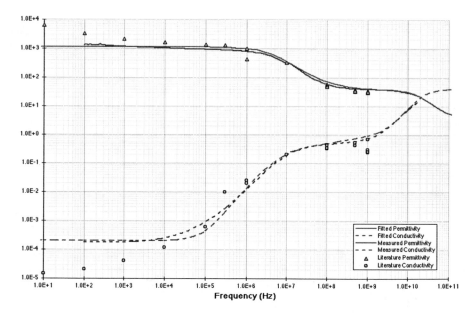

Figure 5.5: Skin (dry) tissue (▲● Literature: [92]-[99]).

Publication list

Articles in refereed journals

1. S. Nishizawa, H.-O. Ruoß, W. Spreitzer, F. Landstorfer, and O. Hashimoto, Induced Current Density in the Human Body using Equivalent Sources for Low-Frequency inhomogeneous Fields, IEICE Transaction Communication, Vol.E84.C, No.10, pp.1612-1614, October 2001.

2. S. Nishizawa, F. Landstorfer, and O. Hashimoto, Experimental study on equivalent magnetic source in ELF range, IEICE, B, Vol.J86-B, No.7, pp.1251-1254, July 2003 (in Japanese).

3. S. Nishizawa, H.-O. Ruoß, F. Landstorfer, and O. Hashimoto, Numerical study on an equivalent source model for inhomogeneous magnetic field dosimetry in the low frequency range, IEEE Transactions on Biomedical Engineering, Vol. 51, No.4, pp.612-616, April 2004.

4. S. Nishizawa, F. Landstorfer, and O. Hashimoto, Study of the magnetic field properties around household appliances using coil source model as prescribed by the European standard EN50366, IEICE Transaction Communication, Vol.E87-C, No.9, pp.1636-1639, September 2004.

Articles in international proceedings

1. S. Nishizawa, H.-O. Ruoß, W. Spreitzer, and F. Landstorfer, A study of induced current density of the human body by using the equivalent sources for inhomogeneous ELF magnetic fields, Proceeding of Asia Pacific Radio Science Conference 2001, PK3-15, p.409, Tokio Japan, August 2001.

2. S. Nishizawa, W. Spreitzer, H.-O. Ruoß, and F. Landstorfer, Equivalent Source Model for Electrical Appliances emitting low Frequency Magnetic Fields, Proceeding of 31th European Microwave Conference 2001, Vol.3, pp.117-120, London UK, September 2001.

3. S. Nishizawa, F. Landstorfer, and O. Hashimoto, Equivalent Source Model for ELF Dosimetry emitted by real household appliances, Proceeding of 25th Biolelectromagnetics Society (BEMS), P-46A, p.213, Maui USA, June 2003.

4. S. Nishizawa, F. Landstorfer, and O. Hashimoto, Induced current densities in an anatomical human body model caused by emission from household appliance, Proceeding of 6th European Conference on Wireless Technology, pp. 133-135, München Germany, October 2003.

5. S. Nishizawa, N. Angwafo, H.-O. Ruoß, W. Spreitzer, F. Landstorfer, and O. Hashimoto, Calculation of current densities induced in an anatomical model of the human body caused by emission from household appliances in the low frequency range, Proceeding of 6th International Congress of the European Bioelectromagnetics Association (EBEA), O-5-07, p.75, Budapest Hungary, November 2003.

6. S. Nishizawa, and F. Landstorfer, Dosimetric Study with household Appliances using the coil source model as prescribed by the European Standard EN50366, Proceeding of 36th Nichtionisierende Strahlung Sicherheit und Gesundheit (NIR 2004), Band 1, Kapitel 11 Messung und Bewertung 3, pp.486-498, Köln Germany, September 2004.

7. S. Nishizawa, F. Landstorfer, and O. Hashimoto, Dosimetric study of induction heater using the coil source model prescribed by the EN50366, Proceeding of 3rd International Workshop on Biological Effects of Electromagnetic Fields, Volume 2, pp.894-903, Kos Greece, October 2004.

Articles in national proceedings

1. S. Nishizawa, H.-O. Ruoß, W. Spreitzer, F. Landstorfer, and O. Hashimoto, Induced Current Density in the Human Body using Equivalent Sources for Low-Frequency inhomogeneous Fields, Proceeding of the 2001 Communications Society Conference of IEICE, B-4-4, Tokio Japan, September 2001.

2. S. Nishizawa, W. Spreitzer, F. Landstorfer, and O. Hashimoto, Study of an equivalent source model using the magnetic dipole moment in ELF range, Technical Report of IEE, EMC-01-15, pp.19-23, Osaka Japan, December 2001 (in Japanese).

3. S. Nishizawa, W. Spreitzer, F. Landstorfer, and O. Hashimoto, Validation of an equivalent source model for measured real household appliances, Proceeding of the 2002 IEICE General Conference, B4-4-74, Tokio Japan, March 2002.

4. S. Nishizawa, F. Landstorfer, and O. Hashimoto, Measurement method and dosimetric evaluation in electric power field using the equivalent source model, Proceeding of the 2003 IEEJ, S6-8, Tokio Japan, March 2003 (in Japanese).

5. S. Nishizawa, W. Spreitzer, and F. Landstorfer, Calculation of current densities induced in an anatomical model of the human body caused by emission from household appliances in the low frequency range, Proceeding of Kleinheubachertagung, KH2003-A-00037, p.34, Kleinheubach Germany, October 2003.

6. S. Nishizawa, Y. Kamimura, F. Landstorfer, and O. Hashimoto, Study of dosimetric calculation of anatomical human body model caused by emission from hand mixer in the ELF range, Proceeding of the 2004 IEICE General Conference, B-4-7, Tokio Japan, March 2004.

7. S. Nishizawa, F. Landstorfer, and O. Hashimoto, Study of the magnetic field properties around household appliances using magnetic source models as prescribed by the European standard EN50366, Proceeding of the 2004 IEEJ General Conference, D211-B2, Tokio Japan, March 2004.

8. S. Nishizawa, F. Landstorfer, Y. Kamimura, O. Hashimoto, T. Tanaka, and T. Noda, Reduction of leakage magnetic fields around the induction cook-

ing heater using the equivalent source model, Proceeding of the 2005 IEICE General Conference, B4-28, Osaka Japan, March 2005 (in Japanese).

Curriculum Vitae

Shinichiro Nishizawa, Ph.D.

born on 17. June 1972 in Tokio, Japan.

1979 - 1982 Elementary School (Albert Schweitzer Schule), Dudweiler-Saarland,
Germany.

1982 - 1984 Elementary School (Deutsche Schule in Tokio), Tokio, Japan.

1984 - 1985 Elementary School (Tokio Gakugei Daigaku Fuzoku Oizumi), Tokio, Japan.

1985 - 1988 Junior high school (Keimei Gakuen), Tokio, Japan.

1988 - 1991 Senior high school (Keimei Gakuen), Tokio, Japan.

1991 - 1995 Aoyama Gakuin University, Department of Electrical Engineering and Electronic,
Tokio, Japan.
Degree : BS (Bachelor of Science)

1995 - 1997 Aoyama Gakuin University, Department of Electrical Engineering and Electronic,
Graduate School of Science and Engineering, Tokio, Japan.
Degree : MS (Master of Science)

1997 - 2000 Aoyama Gakuin University, Department of Electrical Engineering and Electronic,
Graduate School of Science and Engineering, Tokio, Japan.
Degree : Ph.D. (Doctor of Philosophy)

2000 - 2003 Postdoctoral fellowship of Japan Society for the Promotion of Science -
JSPS (PD) at the Aoyama Gakuin University, Department of Electrical Engineering
and Electronic, Tokio, Japan. 2000 and 2002, Guest researcher at the
Institut für Hochfrequenztechnik der Universität Stuttgart, Stuttgart, Germany.

2003 - 2005 Postdoctoral fellowship of Alexander von Humboldt Foundation, at the
Institut für Hochfrequenztechnik der Universität Stuttgart, Stuttgart, Germany.